THE PHOTOGRAPHIC ATLAS OF THE STARS

HJP ARNOLD, P DOHERTY AND P MOORE

with a foreword by Sir Arnold Wolfendale

CRC Press
Taylor & Francis Group
Boca Raton London New York

CRC Press is an imprint of the
Taylor & Francis Group, an **informa** business

First published 1997 by IOP Publishing Ltd.

Published 2019 by CRC Press
Taylor & Francis Group
6000 Broken Sound Parkway NW, Suite 300
Boca Raton, FL 33487-2742

© 1999 by Taylor & Francis Group, LLC
© Photographs: Space Frontiers Limited 1999
CRC Press is an imprint of Taylor & Francis Group, an Informa business

No claim to original U.S. Government works

ISBN 13: 978-0-7503-0654-6 (pbk)
ISBN 13: 978-0-7503-0378-1 (hbk)

Visit the Taylor & Francis Web site at
http://www.taylorandfrancis.com

and the CRC Press Web site at
http://www.crcpress.com

British Cataloguing-in-Publication Data
A catalogue record for this book is available from the British Library

Library of Congress Cataloguing-in-Publication Data are available

Book project managed by John Adamson
Designed by Chris Jones, Design for Art and Science

Typeset in Galliard, Gill Sans and Mantinia by Paul Samat

CONTENTS

To say that there is a wealth of literature about astronomy is to put it mildly – there are books for youngsters, books for amateurs and book after book for professionals. What case can be made for yet another – and for this one in particular? The case is in fact strong, very strong, in that its role is to cater for the gifted amateur – and in my experience the majority of amateurs are very gifted – and the professional who wants a wider view of the heavens than commonly available.

By "wider view" is meant a most impressive collection of star maps and their associated colour photographs. The production of this atlas has represented a labour of considerable proportions: the photographs were taken by Douglas Arnold using carefully controlled techniques in various parts of the world and the processing of the colour film has been carried out with equal care.

To produce coloured images of the stars is no mean task, and of course in principle it is impossible to present images which look the "right colour" to everyone, insofar as perception of colour has a big personal component and, furthermore, viewing conditions can have a big effect. Nevertheless, most observers will be able to appreciate the coloured images and thereby use the atlas for location purposes. In the atlas we see Betelgeux displaying its orange colour and the Pleiades with their bluish cast, the effect of dust in the interstellar medium; these colours are faithfully reproduced in the atlas.

The present writer has recently returned from Australia where he again made the acquaintance of the magnificent southern skies. The glorious images of the region containing the Southern Cross and the Sagittarius star-clouds nicely reproduced in the atlas bring back happy memories.

Accompanying the photographs are the star maps carefully charted by Paul Doherty and a lively text by Patrick Moore; these are also impressive. A nice balance of interesting objects has been struck. We have information about the obviously coloured stars and nebulæ, descriptions of the mythical beasts after which the constellations are named, notes of interest to the historian – such as when a star was first detected, distances and luminosities, and so on.

Amateurs will love this atlas whether as a reference in searches for asteroids or comets in their own photographs or in their own observations. Those interested in the relationship between astronomy and other areas of cosmic science will find the atlas of value too – those searching for the origin of cosmic rays, and anxious to know what interesting phenomena are to be found in their typically 5 degree radius fields, will pore over it!

Sir Arnold Wolfendale FRS
President
The Institute *of* Physics
July 1996

T he reasoning behind the concept of this photographic star atlas is very straightforward. When I became interested in astronomy years ago I took the usual path of learning the sky from planispheres and star maps. It took some time to learn to interpret the stark representation of the stars in the maps in terms of what my eyes were actually seeing in the sky – and there was a further complication. Usually these publications contained images of various objects which frequently were obtained using different instruments and which as a result were not at the same scale. This could be confusing, particularly to a beginner.

Subsequently I thought of a possible solution. Why not secure good quality photographs of the constellations all at the same scale – that is, taken with the same focal length lens? Furthermore, why not use colour film for the photographs so that there was a very rough approximation of how the eye saw the sky? And it need not stop there. If colour transparency film was used the individual transparencies could be printed on standard black and white paper to yield a reversed out image -- black stars on a white background – which could form the basis of an accompanying star map of the particular scene. The locations of stars and other bodies in the map would thus by definition be accurate (since they were based on the photograph) and the artist could concentrate on accurately plotting right ascension and declination lines, identifying and naming various important objects and so on. It was further envisaged that each photograph and the map derived from it would appear on a two-page spread which could be used during observing sessions. Those sessions would be further

Using a fixed camera, this three-and-a-half hour exposure shows the region of the sky around the southern celestial pole star Sigma Octantis. Colour differences and magnitude variations are clearly discernible. The photograph was taken at the Cederberg Observatory in South Africa.

assisted by an informed analysis of what could be seen in each image. This book is the result and it is pleasing for me to see the extent to which the original thinking has been translated into fact.

Of course, there were some important decisions along the way. After much thought, a focal length of 35mm when using a 35mm camera was chosen for the lens since all but one or two of the largest constellations would fit within the angular frame size of 38° (vertical) and 55° (horizontal). My astrophotography is based on the Nikon 35mm SLR system and the lens chosen was the f/1.4 Nikkor F. The use of conventional lenses for recording point sources of light (the stars) is perhaps the most formidable test they can face and for all exposures the lens was stopped down to f/2.8 to limit off-axis aberrations such as coma. The only departure from the use of the 35mm focal length lens resulted from our decision to include a small number of images which would show an even larger area of the sky and thus record the position of the major constellations relative to one another. For these images a Nikkor F ultra-wide-angle18mm f/3.5 lens, with an angular frame size of over 55° by 83°, was used.

Given the lens selected for the bulk of the images, the film speed chosen depended on the amount of detail that we wished to show in a typical image. I originally thought that the photographs should broadly record what the naked eye could discern – objects down to about magnitude 6. On reflection this seemed too restricted and we eventually decided to aim for what could be seen using basic binoculars of 7×50 power. These show objects down to approximately magnitude 8. Previous experience and tests with currently available transparency films showed that generally the result desired could be achieved using ISO400 film in ten-minute exposures, although this was increased to thirty minutes when using the 18mm lens, which is a relatively "slow" lens for photography of point sources such as stars. And that was the norm adopted throughout. Many astrophotographers now use colour negative film to their satisfaction, but this was rejected on three grounds – what was considered to be the continuing (if narrowing) quality advantage of transparency film; because negative film suffers from the disadvantage of the introduction of a further variable with the need to produce a positive print from the original negative; and – most important of all – because of the utter ease with which the black and

white prints could be produced.

In striving to make individual images as comparable as possible in terms of colour and density the ideal would be:

(a) Photography during identical atmospheric conditions – of "seeing" and transparency

(b) The use of the same, standard equipment, and

(c) The use of the same film type, from the same production batch kept refrigerated until use.

David Malin devised an addition to the fixed-camera method of recording star colours. Although still using a fixed camera, he altered the lens focus during time exposures so that star images became progressively more out of focus – that is, spread over a greater area of film. The intention of this "step focus" technique was that somewhere along the cone of each star image an optimum density colour record would be obtained. This photograph shows Betelgeux in Orion recorded by this technique using a 180mm f/2.8 Nikkor ED lens. ISO200 colour film was exposed for 4 minutes in a sky which, however, as the general colour cast shows, was not free from light pollution.

A moderately priced camera drive unit of a type used for many of the pictures reproduced in this book. Integral optics enable the device to be accurately polar aligned and the small motor is driven by battery power. The direction of the motor can be reversed for use in either hemisphere. As shown here, the drive has a small scope fitted for checking the accuracy of the tracking, with the speed of the motor being slowed or increased by using the hand control unit.

This photograph demonstrates the effect of different exposures when recording the image of a star – in this case Vega in the constellation Lyra. Six exposures were made on the same frame of film at two minute intervals with the lens' f-stop being reduced from f/1.2 to f/8 sequentially. The camera was fixed and the length of each exposure was the same – 4 seconds on ISO400 black and white film. The difference between the first and last exposures (those at f/1.2 and f/8, respectively) is marked.

It was easy to achieve (b). The chosen lens was fitted to two Nikon cameras at different times – a Nikon F4S and a Nikon F3HP. Both were fitted with high magnification right-angle viewfinders. The drive platform did vary according to location but provided the equipment performed satisfactorily this was of limited importance. Despite the use of a wide-angle lens, guiding was often used as a means of achieving optimum sidereal tracking.

However, (a) and (c) presented varying degrees of difficulty. Photography during closely comparable atmospheric conditions is a virtual impossibility. (How comparable is comparable?) Good skies were sought and found in a number of locations in the northern and southern hemispheres – at the Star Hill Inn (New Mexico, USA) and the COAA (the Algarve, Portugal) in the former, and at the Cederberg Observatory, north of Cape Town in South Africa and Alice Springs in Australia in the latter. While this overcame the problem of severe light pollution in my home area of Hampshire, England, it could not overcome variations in general atmospheric conditions. However, for the purposes of this book with its emphasis on wide-angle scenes for mapping purposes, the problem was not thought to be too critical.

The use of same make, same batch film should have been possible but was not achieved in practice simply because of the unexpected length of time it took to secure the comprehensive photographic coverage of all of the constellations, an exercise extending over several years. Kodak ISO400 Ektachrome film was used for most of the images appearing in the book although some changes to the emulsion were introduced part of the way through, particularly with the appearance of the Panther version

(called Lumière in the USA). A small number of the images were exposed on 3M's ScotchChrome Professional ISO400 film and the redder colour balance of this film can be identified readily – such as in the pictures from which Maps 9 and 10 in the Atlas section have been derived.

This discussion of how different films record colour raises a more general point. I wrote above that the aim was to produce colour images which yielded a *rough approximation* (the phrase was used advisedly) of what the human eye sees. Quite often this is realized tolerably well – film will record Vega in Lyra as blue-white whilst Betelgeux in Orion is an orange, just as the eye sees them. But some extended objects such as the Great Nebula in Orion create problems since the peak sensitivity of the human eye causes us to see it as a greenish colour while the response of film renders it as reddish. Furthermore, in time exposures the colours of bright stars tend to be somewhat washed out by overexposure whereas those of less bright stars remain or indeed are too dim to discern. Similarly the apparent size of stars in an image can be distorted. This subject is treated at greater length in the captions to the images, which are also intended to suggest projects for those readers interested in making their own investigations.[1]

One of the best ways of recording an approximation of the different colours (and therefore spectral types) of the stars is one of the simplest. Using a fixed camera – that is one which is not on a platform which moves to compensate for the rotation of the Earth – and with access to a dark sky, stars record on the film frame as trails. Because the light trace is moving across the film, the brightest stars do not overexpose as tends to happen with a tracking camera and a reasonable record of the different star colours is obtained, subject to variations in the characteristics of the various colour films that are available. Star trails are also a useful means of demonstrating celestial geometry.

Today's high speed films and fast lenses enable point source images of the main stars in constellations to be obtained, even with a fixed camera. There is a useful, rough rule of thumb for this form of astrophotography. At the worst case location of the Earth's equator (where stars appear to move for the greatest distance in the shortest time) take the value 500, and at, say, a less demanding latitude of 50°, the value 1000. Divide these values – or estimated values based on the latitude of other locations – by the focal length, expressed in terms of millimetres, of the lens being used. The resulting number is the length in seconds of the exposure that can be given without the image on 35mm film showing signs of trailing. The technique is a useful one for amateurs wishing to secure basic images of the sky without the use of a tracking camera platform. There are penalties, however. Only the brightest stars will be recorded and snatching a picture with the lens wide open at maximum aperture leads to distortion of the stars progressively to the edge of the picture frame as the result of off-axis aberrations.

The original idea for this book was mine but its appearance has depended very much on the talents of my two co-authors. Paul Doherty has developed the information content of the basic black and white images into star maps with his customary elegant skill. Some of

The value of star trails in demonstrating the geometry of the celestial sphere is well demonstrated in this photograph. The region of the night sky between Leo and Virgo is shown with the celestial equator lying about two-thirds of the way up the frame. At the equator, stars appear to move at their fastest and the trail is horizontal. To the north and south, the trails start to curve increasingly towards their respective celestial poles.

his work may not be readily apparent. For example, he has had to compensate for off-axis aberration at the extremities of images taken with a wide-angle lens where each right ascension (RA) and declination (Dec.) line has had to be plotted carefully against the stars and any deviation accounted for. The lines drawn are for Epoch 2000.0. Patrick Moore's deep knowledge has been put to a different use in analysing the images, for – quite apart from a general description of an asterism – he has needed to identify objects such as variable stars whose appearance will vary with the time at which a particular image was obtained. Incidentally, we feel that the quality

This photograph demonstrates how high-speed film enables the photographer to obtain point-source images with a fixed camera. It shows most of the constellation Perseus and extends down to the Pleiades in the constellation Taurus at lower left. Notice how strongly the orange-red of the M class star Rho Persei is recorded. The exposure was for 20 seconds on ISO1600 Ektachrome Panther film using the 35mm wide-angle Nikkor lens at f/1.4. The location was in north-west India at latitude 28°N.

of the reversed out black and white prints is sufficient for them to be of use to amateurs as a reference in checking for the presence of such objects as asteroids and possibly comets in their own observations or images.

I do not regard myself as an astronomical observer but very much as an astrophotographer (or, perhaps in these CCD days, I should say an *astroimager*). But during the production of this book I have learned much from Paul and Patrick and if in turn our combined efforts help others in their quest for the stars then all three of us will be well content.

ACKNOWLEDGEMENTS

Considerable help was extended to me at the locations from which the pictures in this volume were photographed and I am particularly grateful to Willem Hollenbach and Steve Kleyn at the Cederberg Observatory in South Africa; Karl Kramer in Alice Springs, Australia; Bev Ewen-Smith at the COAA in the Algarve, Portugal; and Phil Mahon at the Star Hill Inn, New Mexico, USA. For the photographic laboratory work that the production of the book required, I benefited greatly from the professional skill of John Plimmer, Jane Churcher and Matthew Hunt at Japics in Havant and that of Chris Beech, Rob Hobbs, Martyn Roberts, and Chris Warnes at CPL Limited in Portsmouth. I am indebted to Geoffrey Crawley, Technical Editor of the *British Journal of Photography* for continuing advice and guidance.

HJP Arnold

This picture was taken to show the effect of light pollution in stellar photography. Exposed in my home area of south Hampshire, England, the contrast with the same region of sky (Aquila and Delphinus) in map 42, which was photographed from a dark sky site, is dramatic.

Note
[1] For a full treatment, see *Colours of the Stars* by David Malin and Paul Murdin, Cambridge University Press, 1984. See also, "In Search of Star Colors" by David Malin in *Sky & Telescope*, October 1986 (pp. 326–30) and "The Truth About Star Colors" by Philip C. Steffey in *Sky & Telescope*, September 1992 (pp. 266–73).

INTRODUCTION TO THE STAR MAPS

In the map descriptions, emphasis is placed upon objects of special interest which are easy to identify on the photographs. Each constellation appears in a main map heading only once, where most of the data are provided. Wherever, because of size or the composition of the photograph, other parts of a constellation appear on other maps, cross references are provided, and a complete listing of all occurrences is given on page 216. In each constellation a table is given with details of all stars above apparent magnitude 4; the positions are for Epoch 2000.0, and the details are in agreement with the authoritative Cambridge catalogue. Fainter stars referred to in the text are given with their magnitudes and spectral types: e.g., Zeta Eridani (4.8, A3).

Bright stars are listed in order of magnitude; other objects in order of right ascension. Following common practice, stars brighter than apparent magnitude 1.4 are referred to as first magnitude stars.

For nebular objects, both the Messier numbers (from the Messier catalogue of 1781) and the NGC or New General Catalogue numbers (JLE Dreyer, 1888) are given. We have also used the numbers in the new Caldwell Catalogue (1995). The Messier and Caldwell catalogues are given on pages 204 and 207.

The spectral types of stars are given because these are linked with the colours of the stars; see opposite page.

These maps are intended for use by observers with modest equipment. In the accompanying lists, we have followed definite limits, but it is a mistake to be too rigid; there are some objects of special interest which should not be excluded, and others which show up unexpectedly well in the photographs.

In general, the following limitations have been applied:

Double Stars. Doubles are not listed where the separation is less than 1.5 seconds of arc, or where the secondary component is below magnitude 8.

Variable Stars. The listed variables reach magnitude 7 or brighter at maximum, and have a range of at least 0.5 of a magnitude. (Remember, however, that many variables – particularly Mira stars – become very faint at minimum, and so do not then show up in the photographs.)

Nebular Objects. In general, those in the lists have an integrated magnitude of 8 or brighter, but these values are always somewhat arbitrary, and it is also true that some nebular objects are remarkably elusive photographically. Messier objects and some Caldwell objects have been included even when they are fainter than the usual limit.

- For galaxies, spiral galaxies are denoted by **S**; **Sa** (condensed nuclei, tight spiral arms); **Sb** (looser arms); **Sc** (less condensed nuclei, much looser arms); **SB** denotes barred spirals, again with subdivisions a, b, c.
- **E** denotes elliptical systems, from **E0** (spherical) to **E7** (very flattened). **Irreg.** denotes irregular systems.
- **M** and **C** numbers refer to the Messier and Caldwell catalogues respectively.

Magnitudes in the map descriptions are given to one decimal place, whereas in the tables they are given to two.

The limiting separation for 10×50 binoculars is 2.5 seconds of arc, and the faintest magnitude is 10.5. This is somewhat arbitrary, and much depends on the quality of the optics, the state of the sky and the visual acuity of the observer.

Variable stars are of many kinds. The following notes are intended only as a general guide to the variables found in the maps.

ECLIPSING BINARIES

Algol	Periods 0.2 day to over 27 years. Almost spherical components. Maximum for most of the time.
Beta Lyræ	Periods over 1 day. Ellipsoidal components; variations going on all the time, with alternate deep and shallow minima.

PULSATING VARIABLES

Mira	Long-period late-type giants; spectra M or later; periods 80 to 1,000 days. Amplitudes may exceed 10 magnitudes. Periods and amplitudes vary from cycle to cycle.
Semi-regular	Late-type giants. Periods from 20 days to several years. Amplitudes less than with Mira stars, and the periods are less well-marked.
RR Lyræ	Spectra A to F; periods 0.2 to 1.2 days; all are about 95 times as luminous as the Sun.
RV Tauri	Supergiants, types F to K. Periods 30 to 150 days. R Scuti is the brightest example.
Cepheids	Spectra F to K; periods 1 to 135 days; very regular. There is a link between period and luminosity.
W Virginis	Population II Cepheids; less luminous than classical Cepheids.

	Kappa Pavonis is the brightest example.
Delta Scuti	Types A to F; periods less than 1 day; small amplitude.
Beta Cephei (Beta Canis Majoris)	B-type subgiants. Very small amplitudes; short periods.

ERUPTIVE VARIABLES

Gamma Cassiopeiæ	Shell stars; type B; occasional outbursts. Amplitude usually less than 2 magnitudes.
T Tauri	Very young stars, varying irregularly; amplitudes up to 1 magnitude.
R Coronæ Borealis	Types B to R. Occasional deep minima, due to soot accumulating in the star's atmosphere.
UV Ceti	Flare stars. Dwarfs of types K to M; occasional very rapid brightening.

CATACLYSMIC VARIABLES

SS Cygni or Dwarf novæ	Periods from 10 to 1,000 days, amplitudes from 2 to 9 magnitudes.
U Geminorum	Outbursts last for several days; the intervals between outbursts are not constant.
Novæ	Sudden brightening of a faint star, due to an outburst in the white dwarf component of a binary system.
Recurrent novæ	Stars which have shown more than one nova-like outburst. T Coronæ Borealis is the best-known example.

Supernovæ	Violent outbursts. (1) The total destruction of the white dwarf component of a binary system. (2) The collapse of a very massive star.

Spectral Types

Type	Colour	Surface temperature, °C	Examples	Spectra
W	White	Up to 80,000	Zeta Puppis	Many bright lines. Wolf-Rayet stars
O	White	40,000–35,000	Regor	Bright and dark lines
B	Bluish-white	25,000–12,000	Rigel, Spica	Helium prominent
A	White	10,000–8,000	Sirius, Vega	Hydrogen dominant
F	Yellowish	7,500–6,000	Polaris, Procyon	Yellow hue usually not marked
G	Yellow	giants 5,500–2,400 dwarfs 6,000–5,000	Capella Sun	Metallic lines appear
K	Orange	giants 4,000–3,000 dwarfs 5,000–4,000	Arcturus Tau Ceti	Weak hydrogen lines, strong metallic lines
M	Orange-red	giants 3,400 dwarfs 3,000	Betelgeux Proxima	Many are variable. Band due to molecules
R	Red	2,600	T Lyræ	Remote; appear faint
N	Red	2,500	R Leporis	Many variable examples
S	Very red	2,600	Chi Cygni	Prominent bands of titanium oxide

Each type is subdivided from 0 to 9. Types R and N are now often combined as Type C. A separate class, Q, is used for novæ.

The Greek Alphabet

It may be useful here to list the letters of the Greek alphabet. They are:

α	Alpha
β	Beta
γ	Gamma
δ	Delta
ε	Epsilon
ζ	Zeta
η	Eta
θ	Theta
ι	Iota
κ	Kappa
λ	Lambda
μ	Mu
ν	Nu
ξ	Xi
ο	Omicron
π	Pi
ρ	Rho
σ	Sigma
τ	Tau
υ	Upsilon
φ	Phi
χ	Chi
ψ	Psi
ω	Omega

WHOLE SKY MAP

NORTHERN POLAR SKY

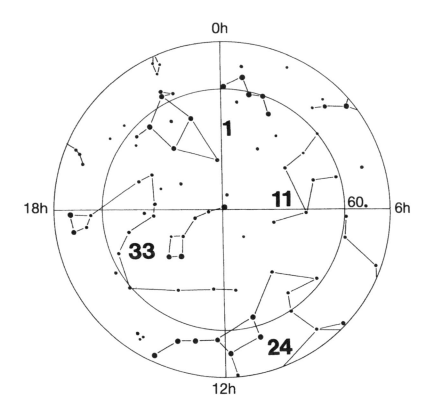

SOUTHERN POLAR SKY

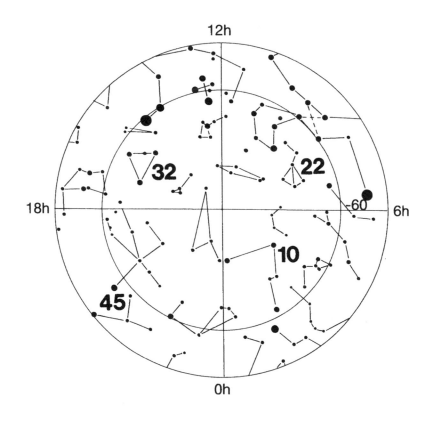

Key to symbols used on the maps

○ open cluster

✦ planetary nebula

✪ globular cluster

☐ gaseous nebula

○ galaxy

∘ variable star

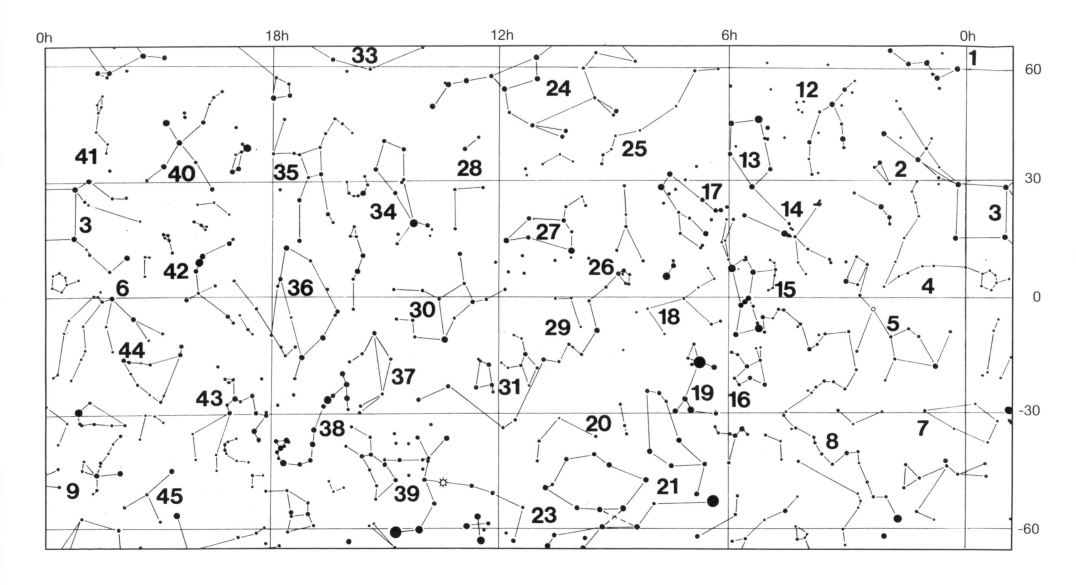

Map A

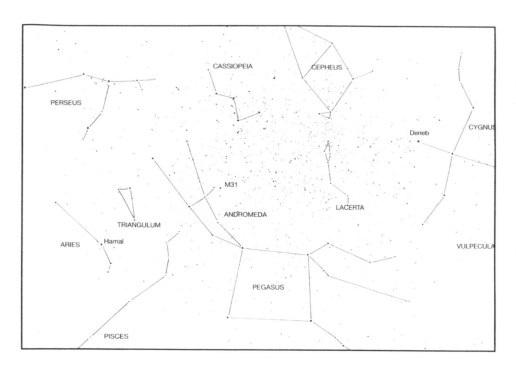

This ultra wide-angle photograph covers a very large and rich area: it contains Perseus, Cassiopeia, Cepheus, Lacerta, Aries, Triangulum, Andromeda, most of Pegasus, and parts of Pisces, Cygnus, and Vulpecula. The only first-magnitude star is Deneb, to the upper right, but there are several stars of the second magnitude. The photograph is crossed by the Milky Way. The W of Cassiopeia is very prominent.

Map B

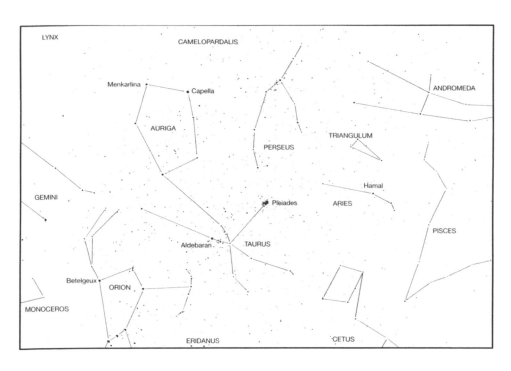

T his ultra wide-angle photograph covers Auriga, Taurus, Perseus, Triangulum, Aries, and parts of Lynx, Camelopardalis, Andromeda, Pisces, Cetus, Eridanus, Orion, Monoceros and Gemini. The Belt of Orion appears at the lower left.

Three first-magnitude stars are included. Of these Capella is obviously the brightest, and its yellow colour is well brought out; compare it with Menkarlina (Beta Aurigæ (1.9, A2)), to the left, which is pure white. When this photograph was taken Betelgeux in Orion was – as usual – decidedly superior to Aldebaran.

Map C

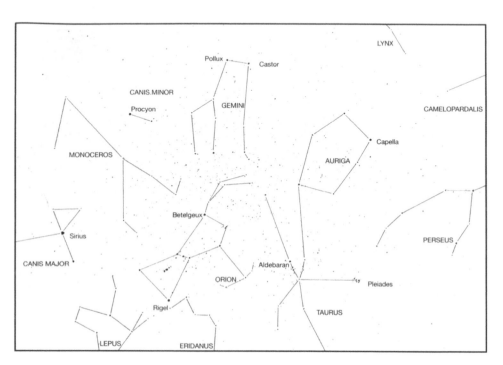

This photograph covers a wide area, and includes Orion, Taurus, Gemini, Canis Minor, Monoceros, and Auriga, together with parts of Perseus, Camelopardalis, Lynx, Eridanus, Lepus and Canis Major. Orion's retinue is well displayed, comprising the first-magnitude stars Sirius, Pollux (with Castor), Procyon, Capella and Aldebaran; the colours of Betelgeux and Aldebaran are very pronounced.

Map D

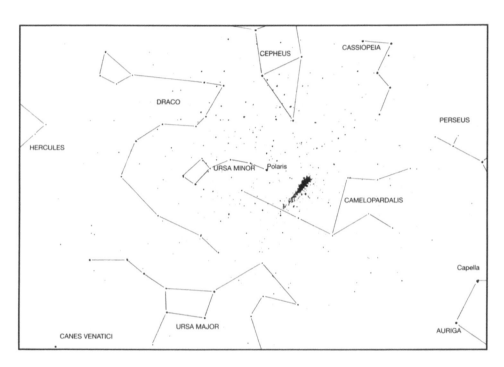

T his ultra wide-angle view shows the north polar region of the sky; Polaris is near the centre, with Ursa Major below and Cassiopeia to the upper right. Capella is seen to the lower right and Perseus near the mid-right edge of the photograph, with Cepheus to the upper right.

The photograph was taken on 27 March 1996. At this time Comet Hyukatake was at its best, and is shown here with its long tail. The orbit is very eccentric, and the period is of the order of 15,000 years.

Map E

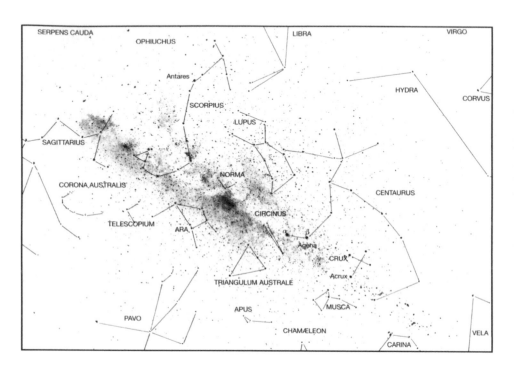

This ultra wide-angle photograph covers what is probably the most glorious region in the entire sky. It extends from Serpens, near the equator, and from Libra and Virgo in the Zodiac, across to Chamæleon and Carina in the far south of the sky. It contains the Scorpion, the Southern Cross and the richest part of the Milky Way, with the magnificent star-clouds in Sagittarius which hide our direct view of that mysterious region, the centre of the Galaxy, almost 30,000 light-years away from us. The photograph gives the impression of looking at an edgewise on galaxy, and it is indeed true that we are looking along the main plane of our system.

Map F

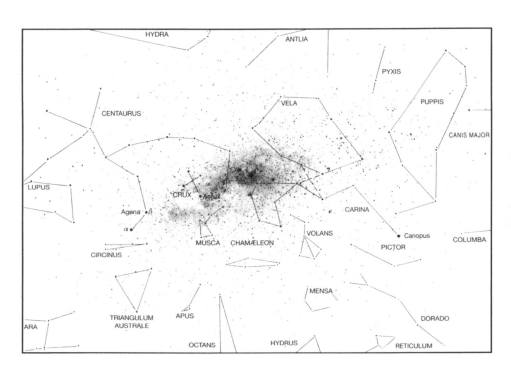

This key map covers a wide area of the southern sky. The regions at the top of the map can be seen from parts of Europe, but at the bottom of the map the observer must "go south".

Of course the photograph is dominated by the richness of the Milky Way in the Carina area; as usual the nebulosity appears red, but this is because the main constituent of the gas is hydrogen, and the redness does not show up to the visual watcher. The Large Cloud of Magellan is very plain to the lower right, on the borders of Dorado and Mensa, and the Coal Sack in the Southern Cross is well marked.

This photograph is particularly well suited for a comparison between the Southern Cross (Crux) and the False Cross (Carina/Vela). The shapes are remarkably similar, though the False Cross is the more symmetrical; also, in each case three of the four stars are white and the fourth orange-red (Gamma Crucis and Epsilon Carinæ respectively). However, Crux is much the smaller and brighter of the two.

Five first-magnitude stars appear; Alpha and Beta Centauri, Alpha and Beta Crucis, and Canopus. The great brilliance of Canopus is well brought out in the photograph.

Map I

CASSIOPEIA, CEPHEUS

This map includes the whole of Cassiopeia and Cepheus, with most of Lacerta and parts of Andromeda, Perseus, Draco and Cygnus.

Cassiopeia is one of the most prominent of the northern constellations, and its W or M pattern, made up of Alpha or Shedir (2.2, variable, K0), Beta (2.3, F2), Gamma (2.2, variable, B0p), Delta (2.7, A5) and Epsilon (3.4, B3) makes it very easy to identify; over Britain and the northern United States and Canada it never sets. Gamma is a "shell star", usually of just below the second magnitude, but capable of producing outbursts which raise it to as high as 1.6. In the photograph it is slightly brighter than Alpha, which is distinctly orange and may itself be variable over a small range (perhaps 2.1 to 2.4). Both Alpha and Gamma may be compared with Beta, which is a normal F-type star – one of our nearer neighbours, at 42 light-years; it is 14 times as luminous as the Sun. Eta and Iota are easy telescopic doubles; both are long-period binaries.

Close to Beta is the remarkable variable Rho Cassiopeiæ. Here it appears equal to the stars to either side of it, Tau (4.9, K1) and Sigma (4.9, B1), but on rare occasions it drops below naked-eye visibility, remaining faint for months before recovering. The spectrum varies between types F and K, so that the colour is slightly yellowish. It is very remote – well over 4,500 light-years away – and may have an absolute magnitude of –8, in which case it is over 200,000 times as luminous as the Sun.

Cassiopeia is crossed by the Milky Way, which runs across the photograph from top left to lower right; the whole area is very rich, and there are several open clusters.

Near Delta lie M103 and NGC 663 (C10); both are sparse, but easy to locate with binoculars. Of more interest is NGC 457 (C13), which is very remote; the star Phi Cassiopeiæ (5.0, F2) lies on the south-east edge of the cluster, and if it is a genuine member it must have an absolute magnitude of about –8. The cluster is certainly very remote; it is clearly shown here, but is more condensed than the other clusters in Cassiopeia and is therefore less easy to resolve into stars. M52 is a rather poor open cluster.

Cepheus adjoins Cassiopeia, but is much less prominent; it is marked by the quadrilateral made up of Alpha, Beta, Iota and Zeta. Gamma, a yellow star of type K1, lies away from the main pattern, and in the far future will become the north pole star.

Forming a triangle with the K-type, rather orange Zeta Cephei are two fainter stars. One, Epsilon (4.2, F0) is a normal Main Sequence star, but the other, Delta, is astronomically one of the most important stars in the sky; it is the prototype Cepheid variable, with a range of from magnitude 3.5 to 4.4 and a period of 5.37 days. The fact that the period of a Cepheid is linked to its real luminosity means that its distance can be calculated – in the case of Delta Cephei itself, about 1,340 light-years. The fluctuations of Delta are easy to follow; suitable comparison stars are Epsilon, Iota, and Nu (4.3, A2).

Nu, roughly between Zeta and Alpha, can also be used as a comparison for the "Garnet Star", Mu Cephei, whose red colour comes out clearly in the photograph, and is striking through binoculars or a telescope; it is said that Mu looks like a glowing coal. Its variations are erratic; the range is between magnitude 3.4 and 5.1 with

a mean of about 4.5. It is a highly luminous star – much more powerful than Betelgeux in Orion – and is well over 1,500 light-years away.

T Cephei, not far from Beta, is a Mira-type variable; when the photograph was taken it was near maximum (magnitude 5.2) but at minimum sinks to below 11. The period is 388 days. Different again is VV Cephei, a huge eclipsing binary near Xi (4.3, F7); the range is from magnitude 4.8 to 5.4, and the red hue of the main M-type component comes out in the photograph. Cepheus contains several open clusters, notably IC 1396, close to Mu, which is well seen with binoculars.

Probably the most interesting object on the map is the Sword-Handle of Perseus, made up of two open clusters, NGC 869 and 884 (C14) (see map 12).

A small part of Andromeda is shown; note the clear yellow colour of 51 Andromedæ, which for some reason was not allotted a Greek letter even though it is the fifth brightest star in the constellation. Its spectral type is K3. The galaxy M31 is described with Map 2.

CASSIOPEIA

Brightest stars

Star		Proper name	Mag.	Spectrum	RA h m s	Dec. ° ' "
γ	Gamma		2.2v	B0p	00 56 42.4	+60 43 00
α	Alpha	Shedir	2.23v	K0	00 40 30.4	+56 32 15
β	Beta	Chaph	2.27	F2	00 09 10.6	+59 08 59
δ	Delta	Ruchbah	2.68	A5	01 25 48.9	+60 14 07
ε	Epsilon	Segin	3.38	B3	01 54 23.6	+63 40 13
η	Eta	Achird	3.44	G0	00 49 05.9	+57 48 58
ζ	Zeta		3.67	B2	00 36 58.2	+53 53 49
ι	Iota		3.98	A1 (combined)	02 03 26.0	+72 25 17

Variable stars

	RA h m	Dec. ° '	Range (mags.)	Type	Period (d)	Spectrum
V	23 11.7	+59 42	6.9–13.4	Mira	228.8	M
ρ	23 54.4	+58 30	4.1–6.2	?		F–K
R	23 58.4	+51 24	4.7–13.5	Mira	430.5	M
T	00 23.2	+55 48	6.9–13.0	Mira	444.8	M
TU	00 26.3	+51 17	6.9–8.1	Cepheid	2.14	F
α	00 40.5	+56 22	?2.1–2.5	Suspected		K
γ	00 56.7	+60 43	1.6–3.3	Irreg.		B
SU	02 52.0	+68 53	5.7–6.2	Cepheid	1.95	F
RZ	02 48.9	+69 38	6.2–7.7	Algol	1.19	A

Double stars

	RA h m	Dec. ° '	PA °	Sep. sec	Mags.	
λ	00 31.8	+54 31	176	0.5	5.3, 5.6	
η	00 49.1	+57 49	293	12.2	3.4, 7.5	Binary, 480 y
ψ	01 25.9	+68 08	113	25.0	4.7, 9.6	
ι	02 29.1	+67 24	232	2.4	4.6, 6.9	Binary, 840 y
σ	23 59.0	+55 45	326	3.0	5.0, 7.1	

Open clusters

M/C	NGC	RA h m	Dec. ° '	Diameter min.	Mag.	No of stars
M52	7654	23 24.2	+61 35	13	6.9	100
	7789	23 57.0	+56 44	16	6.7	300
	129	00 29.9	+60 14	21	6.5	35
C13	457	01 19.1	+58 20	13	6.4	80
	IC 1805	02 32.7	+61 27	22	6.5	40
M103	581	01 33.2	+60 42	6	7.4	25
	1027	02 42.7	+61 33	20	6.7	40
	654	01 44.1	+61 53	5	6.5	60
	659	01 44.2	+60 42	5	7.9	40
C10	663	01 46.0	+61 15	16	7.1	80

Nebulæ

M/C	NGC	RA h m	Dec. ° '	Dimensions min.	Mag. of illum. star
C11 (Bubble Nebula)	7635	23 20.7	+61 12	15 × 8	7
	281	00 52.8	+56 36	35 × 30	8

CEPHEUS

Brightest stars

Star		Proper name	Mag.	Spectrum	RA h m s	Dec. ° ' "
α	Alpha	Alderamin	2.44	A7	21 18 34.6	+62 35 08
γ	Gamma	Alrai	3.21	K1	23 39 20.7	+77 37 57
β	Beta	Alphirk	3.23v	B2	21 28 39.4	+70 33 39
ζ	Zeta		3.35	K1	22 10 51.0	+58 12 05
η	Eta		3.43	K0	20 45 17.2	+61 50 20
ι	Iota		3.52	K1	22 49 40.6	+66 12 02
μ	Mu	"Garnet Star"	3.6v	M2	21 43 30.2	+58 46 48
δ	Delta		3.7v	F8	22 29 10.1	+58 24 55

Variable stars

	RA h m	Dec. ° '	Range (mags.)	Type	Period (d)	Spectrum
T	21 09.5	+68 29	5.2–11.3	Mira	388.1	M
VV	21 56.7	+63 38	4.8–5.4	Eclipsing	7430	M+B
S	21 35.2	+78 37	7.4–12.9	Mira	486.8	N
μ	21 43.5	+58 47	3.4–5.1	Irreg.?		M
δ	22 29.2	+58 25	3.5–4.4	Cepheid	5.37	F–G
W	22 36.5	+58 26	7.0–9.2	Semi-reg.	Long	K–M
U	01 02.3	+81 53	6.7–9.2	Algol	2.49	B+G

Double stars

	RA h m	Dec. ° '	PA °	Sep. sec	Mags.	
κ	20 08.9	+72 43	122	7.4	4.4, 8.4	
β	21 28.7	+70 34	249	13.3	3.2, 7.9	
ξ	22 03.8	+64 38	277	7.7	4.4, 6.5	Binary, 3800 y
δ	22 29.2	+58 25	191	41.0	var., 7.5	
ο	23 18.6	+68 07	220	2.9	4.9, 7.1	Binary, 796 y

Open clusters

M/C	NGC	RA h m	Dec. ° '	Diameter min.	Mag.	No of stars
	IC 1396	21 39.1	+57 30	50	3.5	50
	7160	21 53.7	+62 36	7	6.1	12
	7235	22 12.6	+57 17	4	7.7	30
	7510	23 11.5	+60 34	4	7.9	60
C1	188	00 44.4	+85 20	14	8.1	120

Nebulæ

M/C	NGC	RA h m	Dec. ° '	Dimensions min.	Mag. of illum. star
C4	7023	21 01.8	+68 12	18 × 18	6.8
C9 (Cave Nebula)	Sh2-155	22 56.8	+62 37	50 × 30	7.7

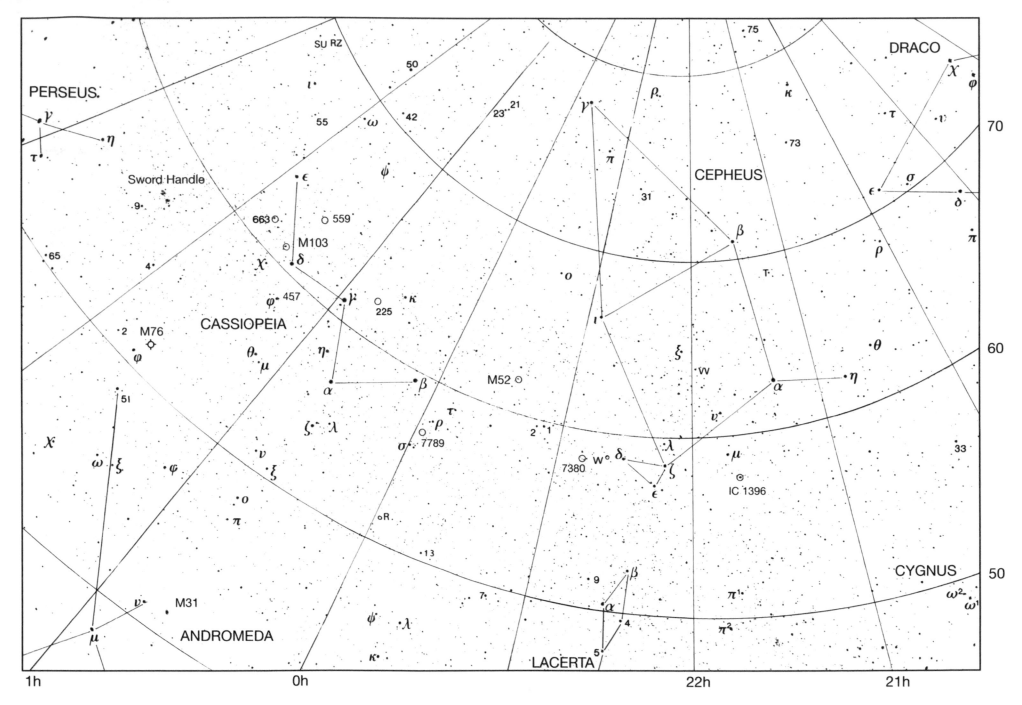

PERSEUS.

DRACO

CEPHEUS

CASSIOPEIA

ANDROMEDA

LACERTA

CYGNUS

Sword Handle

M103
M76
M52
M31

663
559
457
225
7789
7380
7789
IC 1396

SU RZ

1h   0h   22h   21h

70   60   50

Map 2

ANDROMEDA, TRIANGULUM, ARIES

Map 2 contains the whole of Andromeda and Triangulum, most of Aries, parts of Pisces and Pegasus and small portions of Cassiopeia and Lacerta.

The Square of Pegasus is one of the most famous of all the constellations, but for some reason or other the controlling body of world astronomy, the International Astronomical Union, decided to transfer one of its main stars to the adjacent constellation of Andromeda, so that Delta Pegasi (Alpheratz) became Alpha Andromedæ. The whole of Pegasus appears in Map 3, and is described there. Note, however, that in the present map the upper right-hand star of the Square, Beta Pegasi or Scheat, is very obviously orange in colour.

Andromeda is marked by a chain of stars leading off from Alpheratz in the general direction of Perseus; in order they are Delta, Beta and Gamma. Both Beta and Gamma are orange, and appear as such in the photograph; the colour of Beta, an M-type giant, is the more pronounced of the two. Gamma is a fine telescopic double; the components are of magnitudes 2.3 and 4.8; as the separation is almost 10 seconds of arc, a small telescope will split the pair. The secondary component is itself a close binary, with a separation of 0.5 second of arc and a period of 61 years.

There are two very red variables in the constellation. R Andromedæ close to the little triangle made up of Rho (5.2, F5), Theta (4.6, A2) and Sigma (4.5, A2) is a Mira star; it is shown on the photograph, but was then some way from maximum – at its brightest it can rise to magnitude 5.8. W Andromedæ, not far from Gamma, has an S-type spectrum, and the colour comes out well in the photograph, which shows the star when not far from

maximum; at its faintest it drops to below magnitude 14, so that a powerful telescope is needed to show it.

The open cluster NGC 752 (C28) not far from Gamma, is rather sparse, but is not hard to identify when binoculars are used. However, the most celebrated object in Andromeda is of course the Great Spiral, M31, close to Nu (4.5, B5). In the photograph its elongated shape is obvious enough; it is a spiral system, but unfortunately it lies at a narrow angle to us, so that its full beauty is lost (if it were face-on, it would indeed be spectacular). It is visible with the naked eye on a clear night, and is prominent in binoculars, but telescopically it is admittedly a disappointment, and photography is needed to bring out its detailed structure. It is the largest member of the Local Group of galaxies, and lies at a distance of 2.2 million light-years. Its two main companion galaxies, M32 and NGC 205, are dwarf elliptical systems; a very small telescope will show them.

The other important spiral galaxy on this map is M33, in Triangulum. The Triangle is one of the few constellations which really resembles the object after which it is named; the three leaders, Alpha, Beta and Gamma, cannot be mistaken even though they are not bright. Gamma (4.0, A0) lies close to Delta (4.4, G0), and the pair gives the impression of being a wide double, but in fact the two are not genuinely associated; Gamma is 59 light-years from us, Delta only 33.

M33 lies some way from Alpha Trianguli, in the direction of Beta Andromedæ. On the photograph it is just visible as a faint blur. In fact it is fairly easy to see with binoculars, but tends to be very elusive telescopically, because its surface brightness is low. At a distance of

2.9 million light-years it is further away than M31, and it is also smaller, but against this it is presented at a more favourable angle. It is a typical spiral, rather looser than M31.

The red Mira variable R Trianguli lies not far from the Gamma-Delta pair. At maximum it rises to magnitude 5.4 as shown here, but at minimum it falls to below magnitude 12.

Aries, which adjoins Triangulum, used to contain the vernal equinox, which is still known as the First Point of Aries even though the effects of precession have now shifted it into Pisces. The three leaders of Aries make up a fairly obvious little group; Alpha is of type K, and its orange hue contrasts well with the whiteness of Beta. Gamma is a wide, easy double; the components are equal at magnitude 4.6, and the separation is almost 8 seconds of arc. Almost any telescope will show the components separately, though they are too close to be split with binoculars.

Part of Pisces is shown here, including its brightest star, Eta (3.6, G8), but the rest of the constellation appears on Map 4, and is described there. Though Pisces covers a total of 890 square degrees, it is one of the faintest of the Zodiacal constellations.

A small part of Perseus is shown at the top left of the map. Beta Persei (Algol) is the prototype eclipsing binary; below it is Rho, an orange-red semi-regular variable. The open cluster M34, which lies roughly between Algol and Gamma Andromedæ, appears somewhat bluish in the photograph, though the colour is not evident when the cluster is viewed through a telescope. Perseus is shown in its entirety in Map 12, and is described there.

ANDROMEDA

Brightest stars

Star		Proper name	Mag.	Spectrum	RA h m s	Dec. ° ' "
β	Beta	Mirach	2.06	M0	01 09 43.8	+35 37 14
α	Alpha	Alpheratz	2.06	A0	00 08 23.2	+29 05 26
γ	Gamma	Almaak	2.14	K2+A0	02 03 53.9	+42 19 47
δ	Delta		3.27	K3	00 39 19.6	+30 51 40
	51		3.57	K3	01 37 59.5	+48 37 42
o	Omicron		3.6v	B6+A2	23 01 55.1	+42 19 34
λ	Lambda		3.82v	G8	23 37 33.7	+46 27 30
μ	Mu		3.87	A5	00 56 45.1	+38 29 58

Variable stars

	RA h m	Dec. ° '	Range (mags.)	Type	Period (d)	Spectrum
Z	23 33.7	+48 49	8.0–12.4	Z And.		M
KU	00 06.9	+43 05	6.5–10.5	Mira	750	M
R	00 24.0	+38 35	5.8–14.9	Mira	409.3	S
W	02 17.6	+44 18	6.7–14.6	Mira	395.9	S

Double stars

	RA h m	Dec. ° '	PA °	Sep. sec	Mags.	
π	00 36.9	+33 43	173	35.9	4.4, 8.6	
γ {	02 03.9	+42 20	063	9.8	2.3, 4.8	
γ² {			106	0.5	5.5, 6.3	Binary, 61 y

Open clusters

M/C	NGC	RA h m	Dec. ° '	Diameter min.	Mag.	No of stars
	7686	23 30.2	+49 08	15	5.6	20
C28	752	01 57.8	+37 41	50	5.7	60

Galaxies

M/C	NGC	RA h m	Dec. ° '	Mag.	Dimensions min.	Type
	205	00 40.4	+41 41	8.0	17.4 × 9.8	E6
M31	224	00 42.7	+41 16	3.5	178 × 63	Sb
M32	221	00 42.7	+40 52	8.2	7.6 × 5.8	E2

NGC 205 and M32 are companions to M31.

TRIANGULUM

Brightest stars

Star		Proper name	Mag.	Spectrum	RA h m s	Dec. ° ' "
β	Beta		3.00	A5	02 09 32.5	+34 59 14
α	Alpha	Rasalmothallah	3.41	F6	01 53 04.8	+29 34 44

Variable star

	RA h m	Dec. ° '	Range (mags.)	Type	Period (d)	Spectrum
R	02 37.0	+34 16	5.4–12.6	Mira	266.5	M

Double stars

	RA h m	Dec. ° '	PA °	Sep. sec	Mags.
6	02 12.4	+30 18	071	3.9	5.3, 6.9
ι	02 15.9	+33 21	240	2.3	5.4, 7.0

Galaxy

M/C	NGC	RA h m	Dec. ° '	Mag.	Dimensions min.	Type
33	598	01 33.9	+30 39	5.7	62 × 39	Sc

ARIES

Brightest stars

Star		Proper name	Mag.	Spectrum	RA h m s	Dec. ° ' "
α	Alpha	Hamal	2.00	K2	02 07 10.3	+23 27 45
β	Beta	Sheratan	2.64	A5	01 54 38.3	+20 48 29
	c (41)	Nair al Butain	3.63	B8	02 49 58.9	+27 15 38
γ	Gamma	Mesartim	3.9	A0+B9	01 53 31.7	+19 17 45

Double stars

	RA h m	Dec. ° '	PA °	Sep. sec	Mags.
γ	01 53.5	+19 18	000	7.8	4.8, 4.8
π	02 49.3	+17 28	AB 120	3.2	5.2, 8.7
			AC 110	25.2	10.8
ε	02 59.2	+21 20	191	1.5	5.2, 5.5

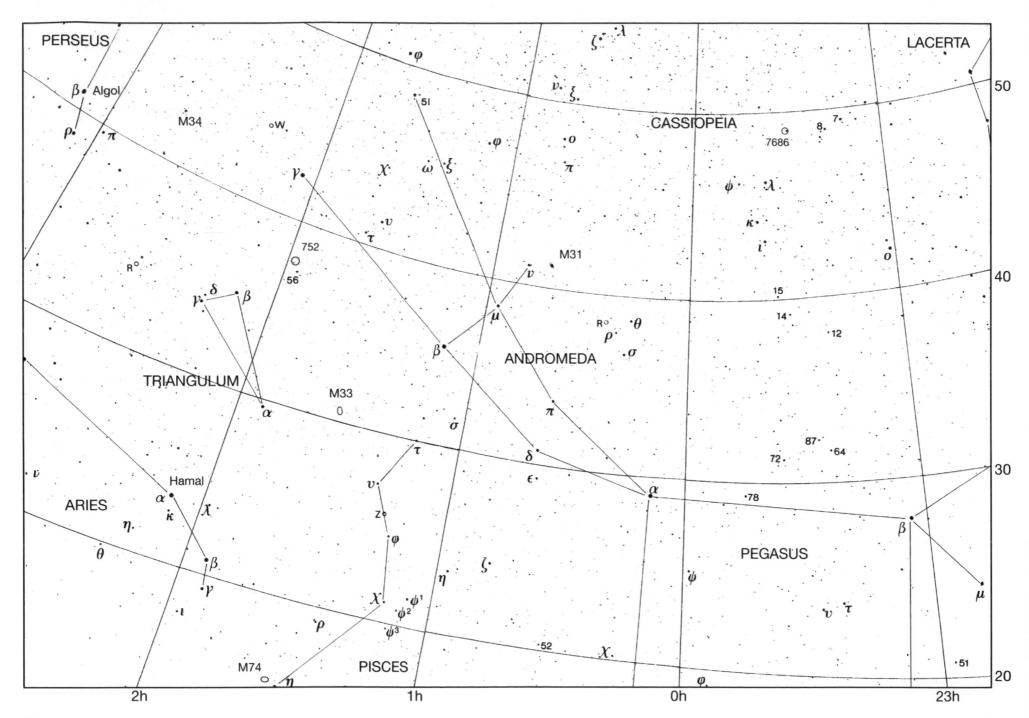

Map 3

PEGASUS

Most of this map is occupied by Pegasus, which covers an area of over 1,100 square degrees and is one of the largest constellations in the sky – though by no means the brightest! Its four main stars make up the famous Square, though, as we have noted, one of the four – Alpheratz, to the upper left in the photograph – has been transferred to Andromeda, as Alpha Andromedæ. Maps often make the Square seem smaller and brighter than it really is, but certainly it is unmistakable.

Three of the four stars of the Square are white, but the fourth, Scheat or Beta Pegasi, is orange, as is well shown in the photograph. It is of type M, though it is not a supergiant such as Betelgeux and is "only" about 300 times as luminous as the Sun; its distance is 176 light-years. Like so many M-type stars, it is variable. The extreme range is from magnitude 2.4 to 2.8, and comparisons may be made with Alpha Pegasi (2.49) and Gamma (2.83, actually an eclipsing binary with a range too small to be detectable with the naked eye). When this picture was taken, Scheat was not far from maximum. It is classed as a semi-regular star; there is a period of 38 days, which is, however, subject to marked irregularities. In colour, Scheat is similar to Beta Andromedæ, seen in the top left corner of the photograph.

It is interesting to see how many stars can be seen within the Square with the naked eye. Of course, binoculars bring out a great many. The brightest are Upsilon (4.4, F8); Tau (4.6, A5); Psi (4.7, M3) and Phi (5.2, M3). Both the latter are red, and despite their faintness their colours are easy to see. Binoculars bring out the hues well. Chi Pegasi, outside the Square and roughly between Alpheratz and Gamma Pegasi, is another red star (4.8, M2).

51 Pegasi, just outside the Square, is of special interest. It is of magnitude 5.5, with a G4-type spectrum; it is 42 light-years away, and in luminosity it is just about equal to the Sun. In 1955 M Mayer and D Queloz, using the 185-centimetre (74-inch) telescope at the Haute-Provence Observatory in France, claimed that they had measured changes in radial velocity which indicated the presence of a planet moving round the star. They gave the mass of the planet as about half that of Jupiter, with a very short revolution period of 4.2 days – in which case the planet would be no more than 6.7 million kilometres (4.2 million miles) from the star, and would be heated to a temperature of at least 1,000 degrees Celsius. This would certainly be surprising, and the discovery has yet to be confirmed. The position of 51 Pegasi is RA 22h 57m 27.8s, dec. +2° 46' 08".

The other bright star of Pegasus is Epsilon, well to the right of the Square. It is of type K, and obviously orange. It has been suspected of variability, though even if this is correct the range must be very small.

Two fainter variables in Pegasus are worth noting. TW is a semi-regular, of type M and therefore red; when this photograph was taken the star was not far from maximum (magnitude 7.2) and is clearly shown. R Pegasi is a Mira variable which can peak at magnitude 6.9, but was not recorded here because it was near minimum (below magnitude 13).

Not far from Epsilon is the splendid globular cluster M15, which is not far below naked-eye visibility and is prominent with binoculars; it has a condensed centre, and was discovered as long ago as 1746 by the Italian astronomer Maraldi, who was searching for a comet at the time. It lies almost directly in line with Theta and Epsilon. The barred spiral galaxy NGC 7331 (C30) is much fainter, but can just be seen on the photograph some distance above Eta.

PEGASUS

Brightest stars

Star		Proper name	Mag.	Spectrum	RA h	m	s	Dec. °	'	"
ε	Epsilon	Enif	2.38	K2	21	44	11.0	+09	52	30
β	Beta	Scheat	2.4 max.	M2	23	03	46.3	+28	04	58
α	Alpha	Markab	2.49	B9	23	04	45.5	+15	12	19
γ	Gamma	Algenib	2.83v	B2	00	13	14.1	+15	11	01
η	Eta	Matar	2.94	G2	22	43	00.0	+30	13	17
ζ	Zeta	Homan	3.40	B8.5	22	41	27.6	+10	49	53
μ	Mu	Sadalbari	3.48	K0	22	50	00.0	+24	36	06
θ	Theta	Biham	3.53	A2	22	10	11.8	+06	11	52
ι	Iota		3.76	F5	22	07	00.5	+25	20	42

Variable stars

	RA h	m	Dec. °	'	Range (mags.)	Type	Period (d)	Spectrum
AG	21	51.0	+12	38	6.0–9.4	Z And.	830	WN+M
V	22	01.0	+06	07	7.0–15.0	Mira	302.3	M
TW	22	04.0	+28	21	7.0–9.2	Semi-reg.	956	M
β	23	03.8	+28	05	2.3–2.8	Semi-reg.	38	M
R	23	06.6	+10	33	6.9–13.8	Mira	378	M

Globular cluster

M/C	NGC	RA h	m	Dec. °	'	Diameter min.	Mag.
M15	7078	21	30.0	+12	10	12.3	6.3

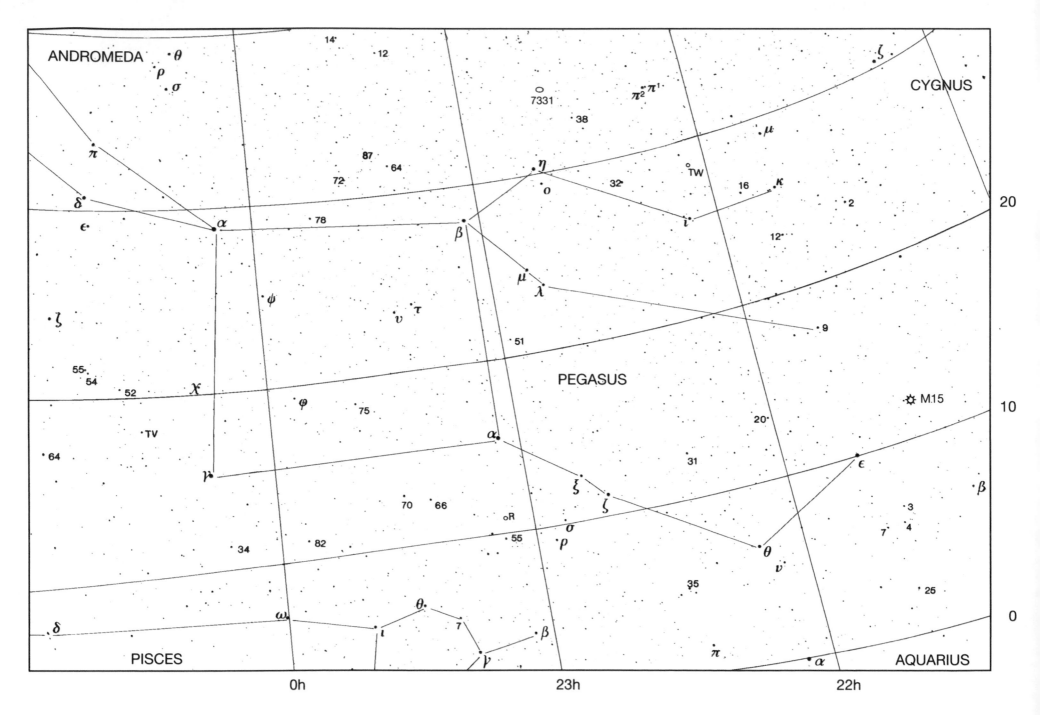

ANDROMEDA

CYGNUS

PEGASUS

PISCES

AQUARIUS

M15

7331

TW

TV

0h

23h

22h

20

10

0

Map 4

PISCES

Here we have a rather barren area of the sky, including almost the whole of Pisces, a part of Cetus, and smaller parts of Aries, Pegasus and Aquarius.

Pisces is one of the least prominent of the Zodiacal constellations, and has only three stars above the fourth magnitude. There are few objects of note, though Alpha is a good telescopic double; the components are of magnitudes 4.2 and 5.1, and the separation is almost 2 seconds of arc (position angle 279°). This is a binary system, with a period of 923 years. Zeta Piscium is a wider double; magnitudes 5.6 and 6.5, separation 23 seconds of arc, position angle 063°. Close to Eta, which at magnitude 3.6 is actually the brightest star in the constellation, is the galaxy M74, one of the fainter Messier objects and not too easy to see on the photograph. It is a rather loose spiral, and is a difficult binocular object, though a telescope of, say, 7.5-centimetre (3-inch) aperture shows it easily enough.

R Piscium is a Mira star, just shown on the photograph because it was well away from minimum; at its dimmest it falls to magnitude 14.8. A telescope is needed to bring out its red colour.

The little pattern of stars in Pisces made up of Iota (4.1, K7), Gamma (3.7, G8), Kappa (4.9, A2) and Lambda (4.5, A7) is not hard to identify, and close beside it is the N-type variable TX (19) Piscium, one of the reddest of all stars. It is variable over a small range and binoculars bring out its colour well; it is evident in the photograph.

Well south of Pisces it is worth seeking out the group in Aquarius consisting of Psi¹, Psi² and Psi³. It looks almost like a very loose cluster, even though the stars are not genuinely associated with each other.

In Cetus, much the most celebrated object is Omicron (Mira), the prototype long-period variable; it was also the first variable to be identified as such (by Phocylides Holwarda, in 1638). The mean period is 332 days, though this is not absolutely constant from one cycle to another. At some maxima Mira may reach the second magnitude, and very occasionally it has risen to 1.7, though at other maxima it fails to exceed magnitude 4; one never knows! At minimum it drops to magnitude 10, beyond the range of binoculars. Generally it is visible with the naked eye for only a few weeks in every year. In this picture it is shown near the left-hand edge, passing through a rather low maximum of about the fourth magnitude; even so, the red hue can be detected. It is large, perhaps 6 million kilometres (3.7 million miles) in diameter, and at its distance of 94 light-years it is one of the nearest of all red giants. it has a 9.5-magnitude binary companion, but the separation is only 0.3 of a second of arc, so that the companion is not an easy object except with a telescope of fair size; it is variable, and even has a variable star designation – VZ Ceti. Two other red variables, AR Ceti (between Mira and the Zeta-Chi pair) and AD (near Iota) are shown on the photograph, but both are of very small range.

Tau Ceti, of magnitude 3.5, is interesting because it is one of the two nearest stars to be reasonably like the Sun (Epsilon Eridani is the other). Tau Ceti is 11.8 light-years away, with a G8-type spectrum and a luminosity slightly less than half that of the Sun. It seems to be a promising candidate as the centre of a planetary system, and efforts have been made – so far without success – to pick up radio signals from it which could be interpreted as artificial.

The Celestial Zodiac

Conventionally, there are twelve constellations of the Zodiac. The vernal equinox, or first Point of Aries, is defined as a point at which the ecliptic and the celestial equator cross; the Sun reaches this position about March 21 each year (the date is not quite constant, owing to the eccentricities of our calendar). Aries is regarded as the first constellation of the Zodiac. However, the effects of precession have now shifted the vernal equinox out of Aries into Pisces, so that logically it is Pisces which should now be regarded as the first constellation of the Zodiac.

The twelve constellations are:

Aries	the Ram
Taurus	the Bull
Gemini	the Twins
Cancer	the Crab
Leo	the Lion
Virgo	the Virgin
Libra	the Balance
Scorpius	the Scorpion
Sagittarius	the Archer
Capricornus	the Sea Goat
Aquarius	the Water-bearer
Pisces	the Fishes

A thirteenth constellation, Ophiuchus (the Serpent-bearer) intrudes into the Zodiac between Scorpius and Sagittarius, so that planets can – and do – pass through it.

PISCES

Brightest stars

Star	Proper name	Mag.	Spectrum	RA h m s	Dec. ° ' "
η Eta	Alpherg	3.62	G8	01 31 28.9	+15 20 45
γ Gamma		3.69	G8	23 17 09.7	+03 16 56
α Alpha	Al Rischa	3.79	A2	02 02 02.7	+02 45 49

Variable stars

	RA h m	Dec. ° '	Range (mags.)	Type	Period (d)	Spectrum
TX	23 46.4	+03 29	6.9–7.7	Irreg.		N
TV	00 28.0	+17 24	4.6–5.4	Semi-reg.	70	M
Z	01 16.1	+25 46	7.0–7.9	Semi-reg.	144	N
R	01 30.6	+02 53	7.1–14.8	Mira	344	M

Double stars

	RA h m	Dec. ° '	PA °	Sep. sec	Mags.	
ζ	01 13.7	+07 35	063	23.0	5.6, 6.5	
ψ¹	01 05.6	+21 28	AB 159	30.0	5.6, 5.8	
			AC 123	92.6	11.2	
α	02 02.0	+02 46	279	1.9	4.2, 5.1	Binary, 933 y

Galaxy

M/C	NGC	RA h m	Dec. ° '	Mag.	Dimensions min.	Type
M74	628	01 36.7	+15 47	9.2	10.2 × 9.5	Sc

Objects in Cetus are listed with Map 5.

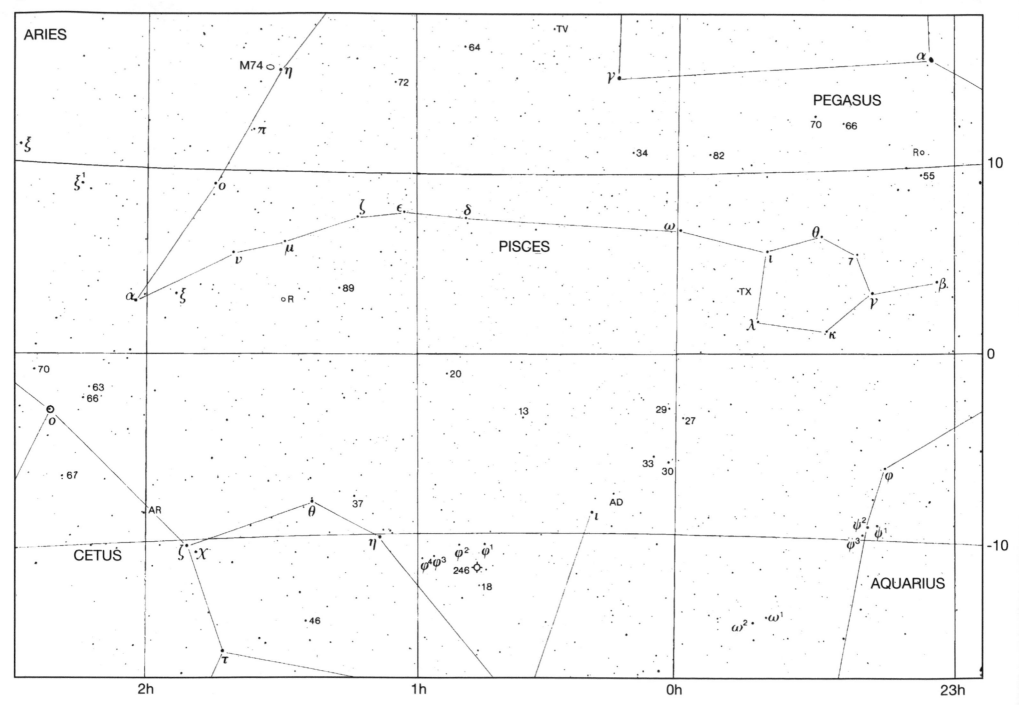

ARIES

PEGASUS

PISCES

CETUS

AQUARIUS

2h   1h   0h   23h

Map 5

CETUS

Here we have the whole of Cetus and much of Pisces, together with parts of Eridanus and Taurus. Pisces has already been described (Map 4).

The first thing to notice is that Mira is much fainter in this photograph than in the photograph with Map 4. The Map 4 picture was taken on October 30 1992 and the Map 5 picture on the following November 25; in the interval Mira has fallen by about one and a half magnitudes, but the red colour is still pronounced. Probably the best way to trace Mira when faint is to identify it first when it is visible with the naked eye, and then memorize the star-fields from a suitable guide such as Delta Ceti (4.1, B2).

The two leaders of Cetus differ in colour. Beta, magnitude 2.0, has a K-type spectrum, but the orange hue is not striking. Alpha (magnitude 2.5) is an M-type giant, and the redness is very evident in the photograph. With Gamma, Mu (4.3, F0), and Xi (4.3, B9) it marks the Whale's head, which is easy to identify even though only Alpha is bright.

Of other variables in Cetus, AD and AR have very small range, and are also shown in Map 4. U Ceti, near Epsilon, was near minimum when the photograph was taken, but at its best can exceed magnitude 7. Another Mira star, R Ceti (near Delta) was closer to maximum, and the colour is evident; it has a period of only 166 days, which for this type of star is rather short. The spectrum is of type M.

UV Ceti, not far from Tau, is a flare star, usually of about magnitude 13. Occasionally it brightens up abruptly, over only a few minutes, and on one occasion reached magnitude 6.8; like all of its kind it is a low-luminosity red dwarf. When this picture was taken, it was not flaring, and is therefore not shown.

The galaxy M77 can just be made out, close to Delta. It was discovered in 1780 by Pierre Méchain, and is a Seyfert system, with a prominent nucleus and weak spiral arms; it is a splendid object in a small telescope. Like most Seyferts it is very active, and is a strong radio source. NGC 247 (C62) is a spiral galaxy south of Beta, and NGC 246 (C56) is a planetary nebula near the little line of stars lettered Phi; it is not easy to see on the photograph, but it is one of the brighter planetaries and a small telescope will show it, though the central star, of magnitude 12, is a difficult object.

The section of Eridanus in this map includes Epsilon, which, like Tau Ceti, is not too unlike the Sun and is close by cosmical standards; its distance is 10.7 light-years, and it has 0.3 the luminosity of the Sun. Whether or not it is attended by planets is not known, but it is at least a promising candidate. Zeta Eridani (4.8, A3) is suspected of having faded from the third magnitude since ancient times. Whether any real secular change has occurred is dubious, because it is not wise to place much reliance upon estimates made centuries ago, but the evidence is rather stronger than with most other similar cases, and it may be significant that despite its current faintness the star was given a proper name – Zibal.

Generally speaking, proper names of stars are used only for the stars conventionally classed as being of the first magnitude (from Sirius down to Regulus) and a few special cases (such as Polaris in Ursa Minor, Mizar in Ursa Major, and Mira in Cetus). The other proper names have fallen into disuse. However, it may be significant that Zeta Eridani was once given a proper name; it has been suggested that this may indicate that it used to be brighter than it is now.

CETUS

Brightest stars

Star		Proper name	Mag.	Spectrum	RA h m s			Dec. ° ' "		
β	Beta	Diphda	2.04	K0	00 43 35.3			−17 59 12		
α	Alpha	Menkar	2.53	M2	03 02 16.7			+04 05 23		
η	Eta		3.45	K2	01 08 35.3			−10 10 56		
γ	Gamma	Alkaffaljidhina	3.47	A2	02 43 18.0			+03 14 09		
τ	Tau		3.50	G8	01 44 04.0			−15 56 15		
ι	Iota	Baten Kaitos Shemali	3.56	K2	00 19 25.6			−08 49 26		
θ	Theta		3.60	K0	01 24 01.3			−08 11 01		
ζ	Zeta	Baten Kaitos	3.73	K2	01 51 27.6			−10 20 06		

Planetary nebula

M/C	NGC	RA h m	Dec. ° '	Dimensions sec.	Mag.	Mag. of central star
C56	246	00 47.0	−11 53	225	8.0	11.9

Galaxies

M/C	NGC	RA h m	Dec. ° '	Mag.	Dimensions min.	Type
C62	247	00 47.1	−20 46	8.9	20.0 × 7.4	S
M77	1068	02 42.7	−00 01	8.8	6.9 × 5.9	SBp

Variable stars

	RA h m	Dec. ° '	Range (mags.)	Type	Period (d)	Spectrum
W	00 02.1	−14 41	7.1–14.8	Mira	351.3	S
T	00 21.8	−20 03	5.0–6.9	Semi-reg.	159	M
S	00 24.1	−09 20	7.6–14.7	Mira	320.5	M
U	02 33.7	−13 09	6.8–13.4	Mira	234.8	M
UV	01 38.8	−17 58	6.8–13.0	Flare		dM
o	02 19.3	−02 59	1.7–10.1	Mira	332	M

Double stars

	RA h m	Dec. ° '	PA °	Sep. sec	Mags.	
37	01 14.4	−07 55	331	49.7	5.2, 8.7	
χ	01 49.6	−10 41	250	183.8	4.9, 6.9	
66	02 12.8	−02 24	AB 234	16.5	5.7, 7.5	
			AC 061	172.7	11.4	
o	02 19.3	−02 59	085	0.3	var., 9.5	(B is VZ Ceti)
ν	02 35.9	+05 36	081	8.1	4.9, 9.5	
ε	02 39.6	−11 52	039	0.1	5.8, 5.8	Binary, 2.7 y
γ	02 43.3	+03 14	294	2.8	3.5, 7.3	

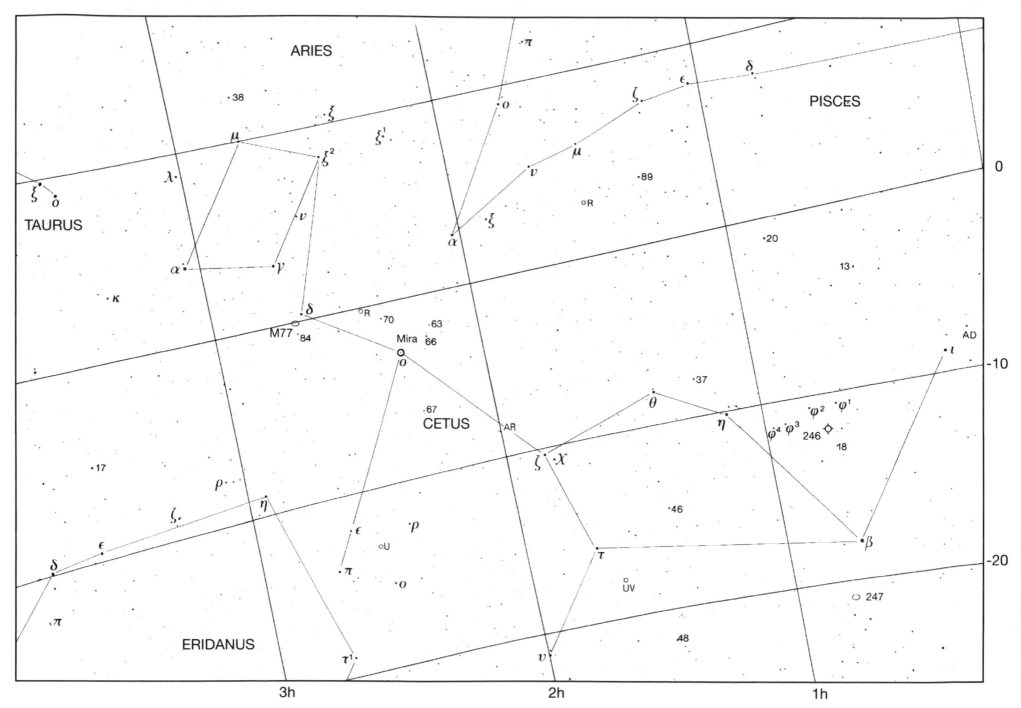

ARIES

PISCES

TAURUS

CETUS

ERIDANUS

3h

2h

1h

38

Map 6

AQUARIUS, EQUULEUS

This map includes most of Aquarius and the whole of Equuleus, together with parts of Pisces and Pegasus. Pisces has been described with Map 4, and Pegasus with Map 3.

Aquarius is a large Zodiacal constellation, but it is not at all prominent, and the only two stars above the third magnitude are Beta (2.9) and Alpha (3.0) – both of which are powerful G-type supergiants, well over 5,000 times as luminous as the Sun and almost 1,000 light-years away. Zeta is a fine binary; the components are of magnitudes 4.3 and 4.5, and the present separation is 2 seconds of arc (position angle 200 degrees). The orbital period is 856 years, so that the separation and position angle change only slowly; the closest separation (1.8 seconds of arc) occurred in 1980, and by 2020 will have widened to almost 3 seconds of arc. Both components are of type F; the distance from us is 98 light-years.

The little group consisting of Phi (4.2, M2); Chi (5.0, M5); Psi[1] (4.2, K0); Psi[2] (4.4, B5) and Psi[3] (5.0, A0) is worth looking at with binoculars, because of the five stars three are red – as is evident in the photograph. Chi is variable over a very small range (4.9 to 5.1).

T Aquarii is a Mira star which can reach magnitude 7.2; it is close to 3 Aquarii (4.4, M3) whose orange hue is well shown here. At the time of the photograph T was well away from maximum, and not easy to detect. Of much greater interest is R Aquarii, close to Omega[1] (5.0, A5) and Omega[2] (4.5, B9). It is what is termed a symbiotic star, of which the prototype is Z Andromedæ. It is a close binary made up of a red giant primary and a bluish subdwarf primary; both seem to be intrinsically variable, so that the overall light-curve is complicated.

It is also erratic, and occasionally the variations are more or less suspended – as in the period from 1931 to 1934, when the magnitude hovered at about 9. The two components are about 1 astronomical unit apart; the distance from us is of the order of 800 light-years. R Aquarii is a source of radio waves, and is associated with faint nebulosity. Owing to its unpredictable behaviour, it is always worth monitoring.

There are two splendid nebular objects in Aquarius. M2 is a globular cluster, more conspicuous than might be thought from the photograph; it is on the fringe of naked-eye visibility, and is easy to see with binoculars, though a telescope is needed to resolve it into stars. It was discovered by Maraldi in 1746, and is 50,000 light-years away, with a real diameter of at least 150 light-years. NGC 7009 (C55) is a planetary, known as the Saturn Nebula; it is about one degree from Nu (4.5, G8) and was discovered by Sir William Herschel in 1782. Its nickname was given to it by Lord Rosse, who observed it in 1850 with his great telescope at Birr Castle, and certainly the shape is slightly reminiscent of the Ringed Planet. The distance is 4,000 light-years, and the real diameter half a light-year. The integrated magnitude is 8.5; the central star is a very hot bluish dwarf with a surface temperature of about 55,000 degrees Celsius. Its magnitude is 11.5, so that with an adequate telescope it is not a difficult object. The other famous planetary nebula in Aquarius, NGC 7293 (C63), is known as the Helix; it is just off this chart and is shown on Map 44 between 41 and 68 Aquarii.

Equuleus is very small and obscure; only Alpha (3.9) is above the fourth magnitude. Alpha makes up a triangle with Delta (4.5, F8) and Gamma (4.7, F0). Delta is a very close binary with equal components and a period of only 5.7 years, but the separation is only 0.3 of a second of arc. There is nothing else of immediate interest in Equuleus.

AQUARIUS

Brightest stars

Star		Proper name	Mag.	Spectrum	RA h m s	Dec. ° ' "
β	Beta	Sadalsuud	2.91	G0	21 31 33.3	−05 34 16
α	Alpha	Sadalmelik	2.96	G2	22 05 46.8	−00 19 11
δ	Delta	Scheat	3.27	A2	22 54 38.8	−15 49 15
ζ	Zeta		3.6	F2+F2	22 28 49.5	−00 01 13
	c² (88)		3.66	K0	23 09 26.6	−21 10 21
λ	Lambda		3.74	M2	22 52 36.6	−07 34 47
ε	Epsilon	Albali	3.77	A1	20 47 40.3	−09 29 45
γ	Gamma	Sadachiba	3.84	A0	22 21 39.2	−01 23 14
	b¹ (98)		3.97	K0	23 22 58.0	−20 06 02

Variable stars

	RA h m	Dec. ° '	Range (mags.)	Type	Period (d)	Spectrum
T	20 49.9	−05 09	7.2–14.2	Mira	202.1	M
R	23 43.8	−15 17	5.8–12.4	Symbiotic	387	M+Pec

Double stars

	RA h m	Dec. ° '	PA °	Sep. sec	Mags.
41	22 14.3	−21 04	AB 114	5.0	5.6, 7.1
			AC 043	212.1	9.0
51	22 24.1	−04 50	AB 324	0.5	6.5, 6.5
			AB+D 191	116.0	10.1
			AC 342	54.4	10.2
			AE 133	132.4	8.6
ζ	22 28.8	−00 01	200	2.0	4.3, 4.5 Binary, 856 y
107	23 46.0	−18 41	136	6.6	5.7, 6.7

Asterism

M/C	NGC	RA h m	Dec. ° '	
M73	6994	20 58.9	−12 38	Four stars; not a true cluster

Globular clusters

M/C	NGC	RA h m	Dec. ° '	Diameter min.	Mag.
M72	6981	20 53.5	−12 32	5.9	9.3
M2	7089	21 33.5	−00 49	12.9	6.5

Planetary nebulæ

M/C	NGC	RA h m	Dec. ° '	Dimensions sec.	Mag.	Mag. of central star
C55 (Saturn Nebula)	7009	21 04.2	−11 22	2.5 × 100	8.3	11.5
C63 (Helix Nebula)	7293	22 29.6	−20 48	770	6.5	13.5

EQUULEUS

Brightest star

Star		Proper name	Mag.	Spectrum	RA h m s	Dec. ° ' "
α	Alpha	Kitalpha	3.92	G0	21 15 49.3	+05 14 52

Variable star

	RA h m	Dec. ° '	Range (mags.)	Type	Period (d)	Spectrum
S	20 57.2	+05 05	8.0–10.1	Algol	3.44	B+F

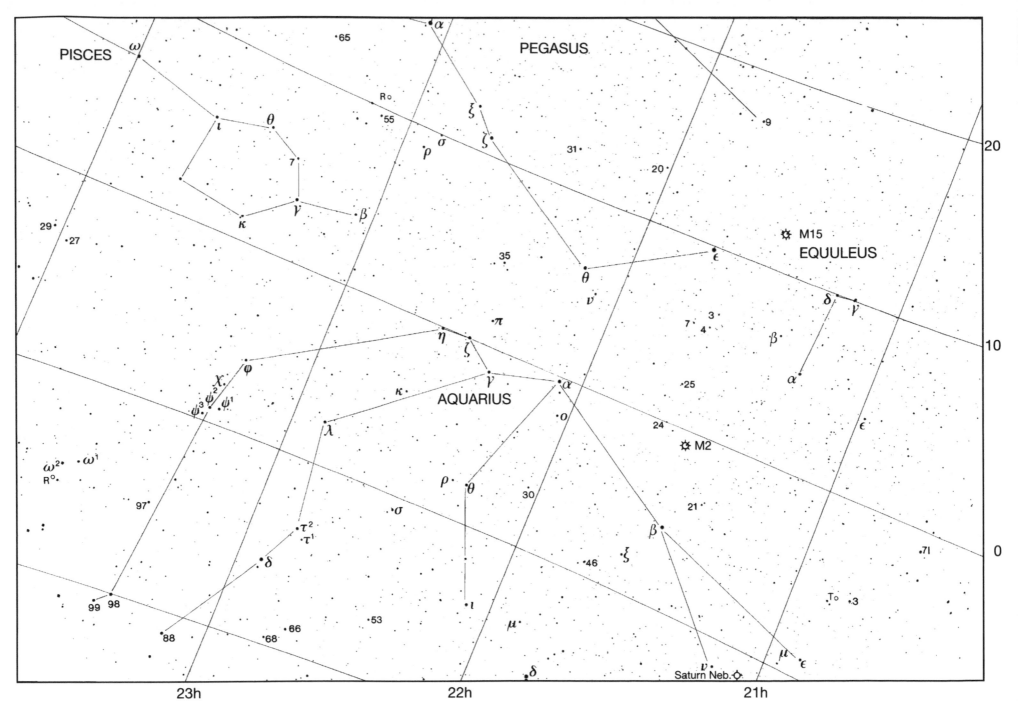

PISCES

PEGASUS

AQUARIUS

EQUULEUS

M15

M2

Saturn Neb.

23h

22h

21h

Map 7

SCULPTOR

Most of this map is taken up with Sculptor, which is a very barren constellation and is notable mainly because it contains the south galactic pole. Much of Piscis Australis is included, with parts of Aquarius, Cetus, Fornax and Eridanus. There are also substantial sections of Phœnix and Grus, but these constellations are better shown in Map 9, and are described there.

Sculptor has no star above the fourth magnitude; its leaders are Alpha (4.3, B8); Gamma (4.4, G8); Beta (4.4, B9) and Delta (4.5, A0). Lambda[1] (6.1, A0) and Lambda[2] (5.9, K0), make up a wide pair, but there is no genuine connection; the fainter star is over 60 light-years further away from us than its orange companion. Epsilon is an easy double.

The red variable S Sculptoris was near maximum (mag. 5.5) at the time of the photograph, so that its colour can be distinguished; its period is 365.3 days, so that it reaches its brightest at almost the same day in each year. Another red variable is the semi-regular, N-type R Sculptoris, which has a smaller range (mag. 5.8 to 7.7) and is easy to identify in the photograph.

The Sculptor system is shown on the chart, but is not readily identifiable on the photograph, because its surface brightness is so low. It is a dwarf elliptical galaxy, and has been likened to a large, very dim globular cluster; it is about 280,000 light-years from us, and is 5,000 light-years in diameter. If the total luminosity is of the order of 3,000,000 times that of the Sun, it is less powerful than a star such as Eta Carinæ. Its position is 4 degrees south from Alpha Sculptoris, but telescopic searches for it are likely to prove abortive.

On the other hand, the famed Sculptor galaxy NGC 253 (C65) is very easily found with a small telescope, and is a splendid example of a spiral seen almost edgewise. It is near the border of Cetus, and about 7.5 degrees south of Beta Ceti, which is shown at the centre top of the photograph (note its orange-yellow hue). In fact the galaxy is more prominent than might be thought from its appearance in the photograph. It is not far outside the Local Group, and is a fairly strong radio source. Close to it in the sky is the globular cluster NGC 288, which is not hard to locate even though it does not show up well here. On the opposite side of Sculptor, near the border with Phœnix, is the galaxy NGC 55 (C72), which has an integrated magnitude of just below 8; it too lies not far beyond the Local Group. Yet another fairly bright spiral is NGC 300 (C70).

Sculptor has one claim to distinction; it includes the south galactic pole. The galactic plane is defined as the great circle which includes the galactic centre and the richest parts of the Milky Way; it is inclined to the celestial equator at an angle of about 63 degrees. The galactic poles are points at galactic latitudes 90 degrees north and south; the north galactic pole lies in Coma Berenices, the south galactic pole in Sculptor.

Part of Piscis Australis is shown here, including the first magnitude Fomalhaut. Piscis Australis is described with Map 9.

SCULPTOR
No star above magnitude 4

Variable stars

	RA	Dec.	Range	Type	Period	Spectrum
	h m	° '	(mags.)		(d)	
S	00 15.4	−32 03	5.5–13.6	Mira	365.3	M
R	01 27.0	−32 33	5.8–7.7	Semi-reg.	370	N

Double stars

	RA	Dec.	PA	Sep.	Mags.	
	h m	° '	°	sec		
δ	23 48.9	−28 08	AB 243	3.9	4.5, 11.5	
			AC 297	74.3	9.3	
ζ	00 02.3	−29 43	320	3.0	5.0, 13.0	
ε	01 45.6	−25 03	028	4.7	5.4, 8.6	Binary, 1192 y

Globular cluster

M/C	NGC	RA	Dec.	Diameter	Mag.
		h m	° '	min.	
	288	00 52.8	−26 35	13.8	8.1

Galaxies

M/C	NGC	RA	Dec.	Mag.	Dimensions	Type
		h m	° '		min.	
	IC 5332	23 34.5	−36 06	10.6	6.6 × 5.1	Sd
	7713	23 36.5	−37 56	11.6	4.3 × 2.0	SBd
	7755	23 47.9	−30 31	11.8	3.7 × 3.0	SBd
	7793	23 57.8	−32 35	9.1	9.1 × 6.6	Sd
	24	00 09.9	−24 58	11.5	5.5 × 1.6	Sb
C72	55	00 14.9	−39 11	8.2	32.4 × 6.5	SB
	134	00 30.4	−33 15	10.1	8.1 × 2.6	SBb
C65	253	00 47.6	−25 17	7.1	25.1 × 7.4	Scp
(Silver Coin Galaxy)						
C70	300	00 54.9	−37 41	8.7	20.0 × 14.8	Sd
	613	01 34.3	−29 25	10.0	5.8 × 4.6	SBb

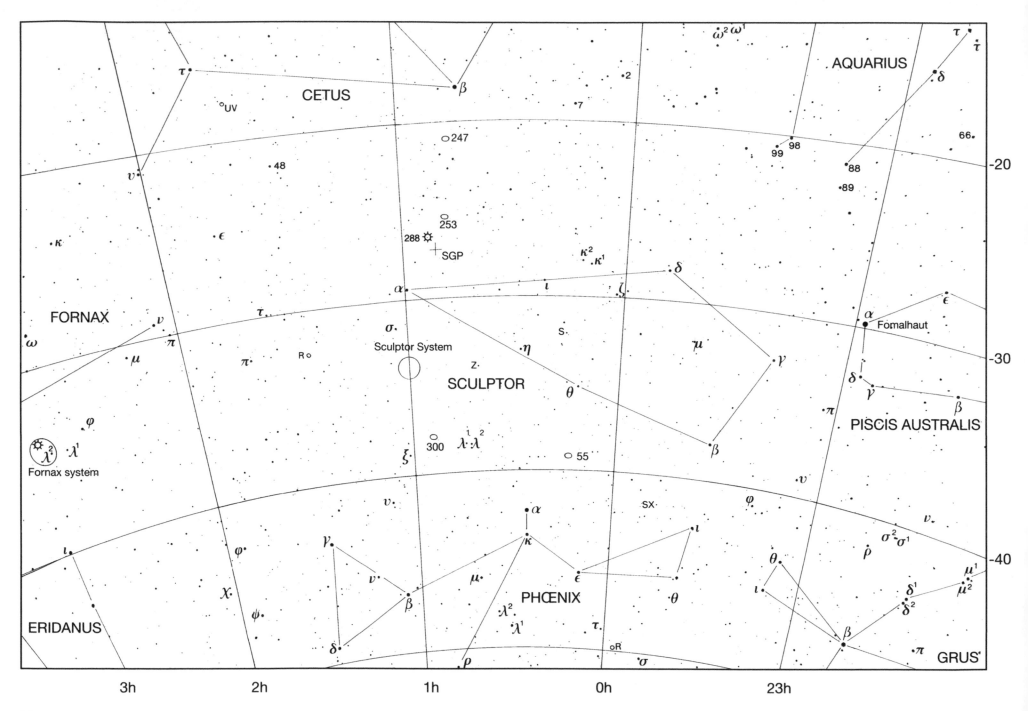

Map 8
ERIDANUS, FORNAX, HOROLOGIUM

Another rather barren region. Much of it is occupied by Eridanus, together with Fornax, Cælum and Horologium; there are also parts of Phœnix and Hydrus, with a section of Dorado. Phœnix is described with Map 9 and Hydrus and Dorado are described with Map 10.

Eridanus is a long, straggling constellation which runs from near Orion to the far south. Its leader, Achernar, is one of the brightest stars in the sky, and is the nearest of the first-magnitude stars to the south celestial pole (it is worth noting that the pole, unmarked by any bright star, lies about midway between Achernar and the Southern Cross). Achernar is invisible from any part of Europe or mainland United States; it just skirts the horizon from Cairo, and is circumpolar anywhere south of Sydney or Cape Town. It is pure white, with a luminosity about 400 times that of the Sun; it is 85 light-years away.

There is a curious puzzle associated with it. The name "Achernar" may be derived from the Arabic *Al Anir Al Nahy*, the "End of the River", but it is by no means certain whether this name originally applied to Alpha Eridani at all. Ptolemy of Alexandria, last of the great astronomers of Classical times (AD 120–180 or thereabouts) did not mention it – which was natural enough, since from Alexandria it never rises. The name may then have been applied to a different star, Theta Eridani, which can be seen from Alexandria because it is further from the pole (declination approximately –40°). So which of the two is the real End of the River? To confuse matters still further, Theta – now known as Acamar – was ranked of the first magnitude by Ptolemy and by the later Arab astronomer Al-Sûfi, but is now only just above the third. It is not the sort of star which would

be expected to show permanent change over the span of a few centuries, and one is bound to wonder whether the original name applied to Alpha after all! Theta, incidentally, is a fine double; the components are of magnitudes 3.4 and 4.5, separated by over 8 seconds of arc (position angle 088°). A small telescope will divide the pair; both are of spectral type A, and their luminosities are respectively 50 and 17 times that of the Sun. The distance from us is 55 light-years.

Iota, near Theta, is clearly orange, and this shows up in the photograph (4.1, K0). Also nearby is S Eridani, which has an F-type spectrum. It was once thought to be variable, but is now regarded as constant at magnitude 4.8. Of the other stars of Eridanus, Tau[4] (3.7, M3) is obviously red.

The galaxy NGC 1291, in the region of Theta, is a spiral with an integrated magnitude of 8.5; it can just be made out in the photograph, but is far from conspicuous. The adjacent galaxy NGC 1316 is actually across the boundary of Fornax, close to the little pair consisting of Chi[1] Fornacis (6.2, A2) and Chi[2] (4.7, K0); the colour difference between the two is clearly discernible in the photograph, and the galaxy can just be made out.

Alpha Fornacis is an easy double, and so is Omega. R Fornacis is a Mira star which can rise to magnitude 7.5; its colour can be seen in the photograph.

The Fornax cluster of galaxies is quite populous, but its leading system, NGC 1316, is only of magnitude 8.8. The nearest fairly identifiable star is Lambda[2] (5.9, K0).

Cælum, which adjoins Eridanus, is entirely unremarkable; it is included in Map 16. Its brightest star, Alpha is only of magnitude 4.4.

Horologium is also very obscure, but it does contain a bright globular cluster NGC 1261 (C87), well within the range of a small telescope. Alpha and Delta make up a wide pair (though they are not genuinely associated) and differ in colour; Alpha is orange, Delta white. R, T and U Horologii are Mira variables. The brightest of them, R, can rise to magnitude 4.7 at maximum, but at the time of this photograph was far from being at its best.

The star Alpha Hydri (2.9, F0) appears at the lower right of the map, below Achernar. The main part of Hydrus is shown in Map 10.

ERIDANUS

Brightest stars

Star	Proper name	Mag.	Spectrum	RA h m s	Dec. ° ' "
α Alpha	Achernar	0.46	B5	01 37 42.9	−57 14 12
β Beta	Kursa	2.79	A3	05 07 50.9	−05 05 11
θ Theta	Acamar	2.92	A3+A2	02 58 15.6	−40 18 17
γ Gamma	Zaurak	2.95	M0	03 58 01.7	−13 30 31
δ Delta	Rana	3.54	K0	03 43 14.8	−09 45 48
υ⁴ Upsilon⁴		3.56	B9	04 17 53.6	−33 47 54
φ Phi		3.56	B8	02 16 30.6	−51 30 44
τ⁴ Tau⁴	Angetenar	3.69	M3	03 19 30.9	−21 45 28
χ Chi		3.70	G5	01 55 57.5	−51 36 32
ε Epsilon		3.73	K2	03 32 55.8	−09 27 30
υ² Upsilon²	Theemini	3.82	K0	04 35 33.0	−30 33 45
53	Sceptrum	3.87	K2	04 38 10.7	−14 18 15
η Eta	Azha	3.89	K1	02 56 25.6	−08 53 54
ν Nu		3.93	B2	04 36 19.1	−03 21 09
υ³ Upsilon³		3.96	M1	04 24 02.1	−34 01 01

Variable star

	RA h m	Dec. ° '	Range (mags.)	Type	Period (d)	Spectrum
Z	02 47.9	−12 28	7.0–8.6	Semi-reg.	80	M

Double stars

	RA h m	Dec. ° '	PA °	Sep. sec	Mags.	
ρ	01 39.8	−56 12	194	11.2	5.5, 5.8	Binary, 484 y
θ	02 58.3	−40 18	088	8.2	3.4, 4.5	

FORNAX

Brightest star

Star	Proper name	Mag.	Spectrum	RA h m s	Dec. ° ' "
α Alpha		3.87	F8	03 12 04.2	−28 59 13

Double stars

	RA h m	Dec. ° '	PA °	Sep. sec	Mags.	
ω	02 33.8	−28 14	244	10.8	5.0, 7.7	
α	03 12.1	−28 59	298	4.0	4.0, 7.0	Binary, 314 y

Galaxies

M/C	NGC	RA h m	Dec. ° '	Mag.	Dimensions min.	Type
C67	1097	02 46.3	−30 17	9.2	9.3 × 6.6	SBb
	1316	03 22.7	−37 12	8.8	7.1 × 5.5	SB0p

HOROLOGIUM

Brightest star

Star	Proper name	Mag.	Spectrum	RA h m s	Dec. ° ' "
α Alpha		3.86	K1	04 14 00.0	−42 17 40

Variable stars

	RA h m	Dec. ° '	Range (mags.)	Type	Period (d)	Spectrum
R	02 53.9	−49 53	4.7–14.3	Mira	404	M
T	03 00.9	−50 39	7.2–13.7	Mira	217.7	M

Globular cluster

M/C	NGC	RA h m	Dec. ° '	Diameter min.	Mag.
C87	1261	03 12.3	−55 13	6.9	8.4

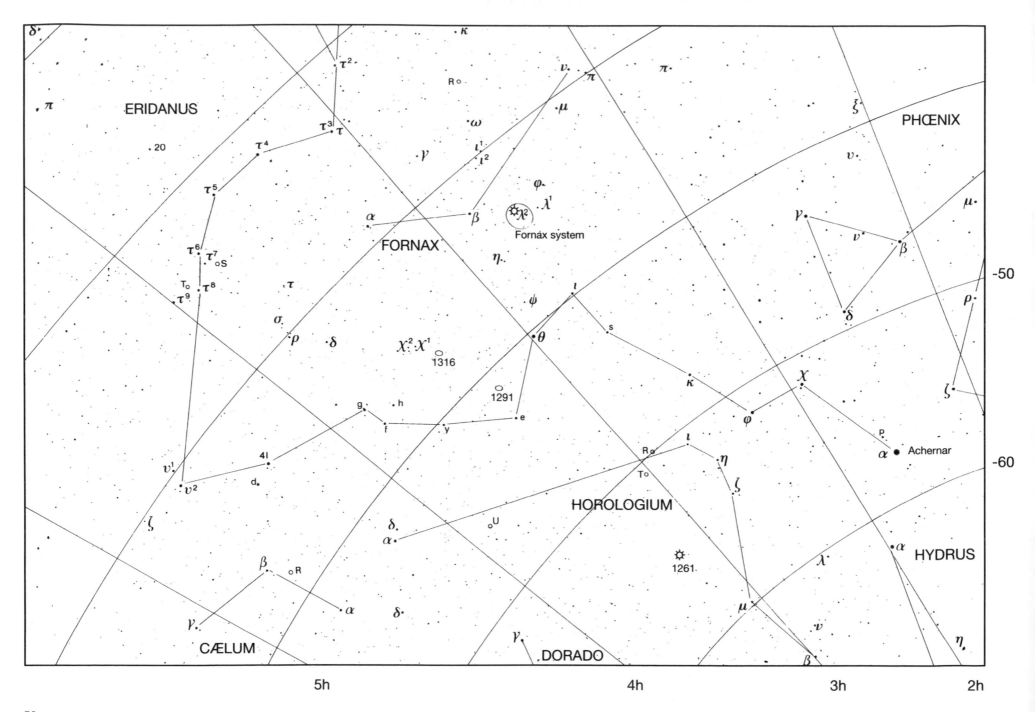

ERIDANUS

PHŒNIX

FORNAX

Fornax system

HOROLOGIUM

CÆLUM

DORADO

HYDRUS

Achernar

-50

-60

5h

4h

3h

2h

50

Map 9

GRUS, PHŒNIX, MICROSCOPIUM, INDUS, PISCIS AUSTRALIS

Here we come to the Southern Birds; the map contains the whole of Grus, Phœnix and Piscis Australis, with sections of Pavo and Tucana. Indus and Microscopium are included, and also Sculptor, which has already been described (Map 7). Fomalhaut appears near the top centre of the photograph, and Alpha Pavonis to the lower right.

Grus is much the most distinctive of the Birds, and it does not need a great effort of the imagination to conjure up the picture of a flying crane. The two leaders make a beautifully contrasting pair. Alpha (Alnair) is a bluish-white B-type star, 23 times as luminous as the Sun and 68 light-years away. Beta (Al Dhanab) is of type M3, and warm orange; it looks slightly fainter than Alnair, but is much more luminous (800 Sun-power) and further away (173 light-years). Binoculars bring out the contrast very well.

There are two pairs in Grus which give the constellation its characteristic appearance: Delta and Mu. The two stars of Delta are not associated; Delta[1] (4.0, G5) is 140 light-years from us, Delta[2] (4.1, M4) only 88 light-years, so that it lies in the foreground, so to speak. Here, too, we have a good colour contrast. The two stars of Mu are G-type giants, and although they are around 7 light-years apart they probably have a common origin. Theta (4.3, F6) and Iota (3.9, K0) mark the Crane's wing; Theta is slightly yellowish, Iota markedly orange. Theta is an easy double.

T and S Gruis, both easily visible on the photograph, are normal Mira variables. Pi[1] is more interesting; when the photograph was taken it was near its maximum of magnitude 5.4, and as it never fades below magnitude 7 it is always within binocular range. As can be seen from the photograph, it is very red; the spectrum is of type S.

Phœnix, the next Bird, is not nearly so distinctive, and Alpha or Ankaa (2.4, K0) is its only bright star; it is of type K, but the orange hue is not very pronounced. It makes up a very wide pair with Kappa (3.9, A3), though there is no true association; Ankaa, at 78 light-years, is much the more distant of the two (it is 60 times as luminous as the Sun). Beta is a good close double with almost equal components.

Zeta Phœnicis is an eclipsing binary of the Algol type. The fluctuations are easy to follow with the naked eye, and there are suitable comparison stars in Eta (4.4, A0); Delta (4.0, K0) and Beta (3.3, G8). Both components of Zeta are of type B. The range is half a magnitude, much less than with Algol itself, but the period is much shorter. In the same binocular field with Iota (4.7, A0) is the remarkable variable SX Phœnicis, which has a range of from 6.8 to 7.5 and a period of only 79 minutes. It is a sub-dwarf, smaller and less luminous than the Sun, and is less than 150 light-years away; it is not unique, but rapidly-pulsating stars of this kind are rather uncommon. It is detectable on the photograph, though it is not too easy to identify. The Mira star R Phœnicis is evident enough; it can reach magnitude 7.5, and like most of its kind it is red in colour.

Microscopium, on the other side of Grus, is very obscure, with no star brighter than Gamma (4.7, G4). Alpha (4.9, G6) has a tenth-magnitude companion, and although this is a wide pair the faintness of the companion makes it rather a difficult object. Gamma was formerly included in Piscis Australis, as 1 Piscis Australis – one of several cases of "transferred stars"; Microscopium itself was added to the sky by the Abbé La Caille in his maps drawn in 1752.

Indus is another faint constellation; its leader, Alpha, is just below the third magnitude. The Mira variable S Indi lies close to Mu (5.2, K5) but was near its minimum at the time of the photograph. Possibly the most interesting star in the constellation is Epsilon (4.7, K5). Its position is RA 21h 59.6m, declination –57° 00'; it is 11.2 light-years away, and has only 13 per cent of the luminosity of the Sun, with an estimated diameter of around 960,000 kilometres (600,000 miles). It lies in the same binocular field with Delta (4.4, F0). Epsilon Indi has the distinction of being the feeblest star visible with the naked eye; its absolute magnitude is +7. It may well be the centre of a system of planets, though of course we have no positive information. It is worth finding, and its orange colour contrasts with the slight yellowish cast of its neighbour Delta. Theta is an easy double.

Piscis Australis (also known as Piscis Austrinus) has only one star above the fourth magnitude; this is Fomalhaut, which at 22 light-years is one of the very closest of the bright stars (of those of the first magnitude

GRUS

Brightest stars

Star		Proper name	Mag.	Spectrum	RA h m s	Dec. ° ' "
α	Alpha	Alnair	1.74	B5	22 08 13.8	−46 57 40
β	Beta	Al Dhanab	2.11	M3	22 42 39.9	−46 53 05
γ	Gamma		3.01	B8	21 53 55.6	−37 21 54
ε	Epsilon		3.49	A2	22 48 33.1	−51 19 01
ι	Iota		3.90	K0	23 10 21.4	−45 14 48
δ¹	Delta¹		3.97	gG5	22 29 15.9	−43 29 45

Variable stars

	RA h m	Dec. ° '	Range (mags.)	Type	Period (d)	Spectrum
R	21 48.5	−46 55	7.4–14.9	Mira	331.9	M
π¹	22 22.7	−45 57	5.4–6.7	Semi-reg.	150	S
S	22 26.1	−48 26	6.0–15.0	Mira	401.4	M

PHŒNIX

Brightest stars

Star		Proper name	Mag.	Spectrum	RA h m s	Dec. ° ' "
α	Alpha	Ankaa	2.39	K0	00 26 17.0	−42 18 22
β	Beta		3.31	G8	01 06 05.0	−46 43 07
γ	Gamma		3.41	K5	01 28 21.9	−43 19 06
ζ	Zeta		3.6 max.	B8	01 08 23.0	−55 14 45
ε	Epsilon		3.88	K0	00 09 24.6	−45 44 51
κ	Kappa		3.94	A3	00 26 12.1	−43 40 48
δ	Delta		3.95	K0	01 31 15.0	−49 04 22

Variable stars

	RA h m	Dec. ° '	Range (mags.)	Type	Period (d)	Spectrum
SX	23 46.5	−41 35	6.8–7.5	Delta Scuti	0.055	A
R	23 56.5	−49 47	7.5–14.4	Mira	267.9	M
S	23 53.1	−56 35	7.4–8.2	Semi-reg.	141	M
ζ	01 08.4	−55 15	3.6–4.1	Algol	1.67	B+B

Double star

	RA h m	Dec. ° '	PA °	Sep. sec	Mags.
β	01 06.1	−46 43	346	1.4	4.0, 4.2

MICROSCOPIUM

No star above magnitude 4

Variable stars

	RA h m	Dec. ° '	Range (mags.)	Type	Period (d)	Spectrum
T	20 27.9	−28 16	7.7–9.6	Semi-reg.	344	M
U	20 29.2	−40 25	7.0–14.4	Mira	334.2	M
S	21 26.7	−29 51	7.8–14.3	Mira	208.9	M

Double star

	RA h m	Dec. ° '	PA °	Sep. sec	Mags.
α	20 50.0	−33 47	166	20.5	5.0, 10.0

INDUS

Brightest stars

Star		Proper name	Mag.	Spectrum	RA h m s	Dec. ° ' "
α	Alpha	Persian	3.11	K0	20 37 33.9	−47 17 29
β	Beta		3.65	K0	20 54 48.5	−58 27 15

Double star

	RA h m	Dec. ° '	PA °	Sep. sec	Mags.
θ	21 19.9	−53 27	275	6.0	4.5, 7.0

PISCIS AUSTRALIS

Brightest star

Star		Proper name	Mag.	Spectrum	RA h m s	Dec. ° ' "
α	Alpha	Fomalhaut	1.16	A3	22 57 38.9	−29 37 20

Double stars

	RA h m	Dec. ° '	PA °	Sep. sec	Mags.	
η	22 00.8	−28 27	115	1.7	5.8, 6.8	
β	22 31.5	−32 21	172	30.3	4.4, 7.9	(optical)
γ	22 52.5	−32 53	262	4.2	4.5, 8.0	
δ	22 55.9	−32 32	244	5.0	4.2, 9.2	

only Alpha Centauri, Sirius, Procyon and Altair are nearer). In 1983 the infra-red astronomical satellite IRAS discovered that it, like Vega, is associated with a cloud of cool material which just may be planet-forming. Indeed, as a potential planetary centre Fomalhaut is not too unpromising. It is hotter than the Sun, with an A-type spectrum, and has 13 times the Sun's luminosity, but there seems no valid reason why it should be definitely ruled out.

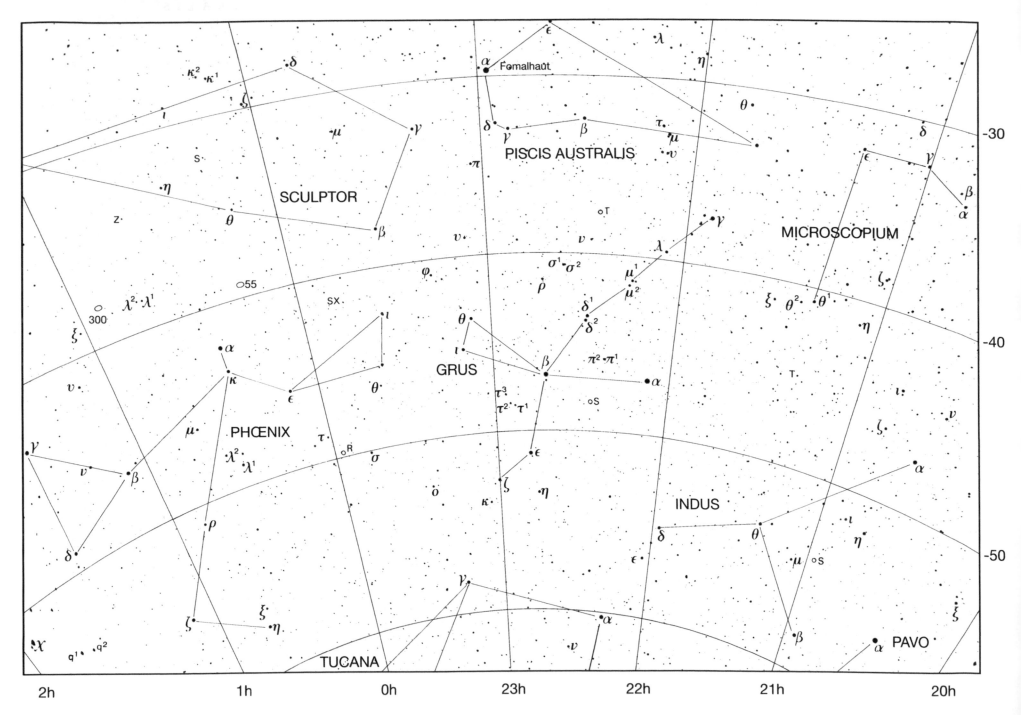

Map 10

DORADO, RETICULUM, TUCANA, OCTANS, MENSA, HYDRUS

This map brings us to the south celestial pole, which is a very barren area. The chart is filled by dim, "modern" constellations: Dorado, Reticulum, Hydrus, Mensa, Pictor, Chamæleon, Volans, most of Octans (the polar constellation) and Tucana, and parts of Eridanus, Phœnix, Horologium and Pavo, with a smaller section of Carina. Canopus appears at the extreme upper left of the photograph, and Achernar at the upper right.

Dorado is shown in part, though its brightest star, Alpha (3.3, A0) is just off the top of the map. The Mira variable R Doradûs is near the border of Reticulum; at the time of the photograph it was near its maximum of magnitude 4.8, and its redness is evident. Close beside it is another Mira star, R Reticuli, but this is not well shown on the photograph because it was not far from its minimum brightness; at maximum it can rise to magnitude 6.5. The "guide star" is Alpha Reticuli, of magnitude 3.3, close to which are Eta (5.1, G7) and Theta (6.2, G9). it is slightly confusing to have two adjacent variables each lettered R! A much brighter variable is Beta Doradûs, which is a classical Cepheid with a range of from magnitude 3.7 to 4.1. Beta Doradûs is very powerful, and according to the Cambridge catalogue its luminosity approaches 200,000 times that of the Sun, in which case it is the equal of Canopus. It is around 2,300 light-years away.

Most of the Large Magellanic Cloud (LMC) lies in Dorado. This is a companion or satellite galaxy to our own system, and lies at a distance of 169,000 light-years;

it is named in honour of the Portuguese explorer Ferdinand Magellan, but it must have been known long before Magellan's time, since it is so bright that not even strong moonlight will drown it. It was formerly classed as irregular, but there are unmistakable signs of a barred spiral form. It is of course much smaller than our Galaxy – about one-quarter the size – but it still ranks as a major system. It contains objects of all kinds, such as variable stars, nebulæ, open and globular clusters, novæ and a vast amount of interstellar material.

In the LMC is the magnificent Tarantula Nebula, NGC 2070 Doradûs (C103); if it were as close to us as the nebula in Orion's Sword, it would cast shadows. It is well shown in the photograph, though it looks rather like a diffuse red star. Also in the LMC is the variable S Doradûs, which is not shown in the photograph but which is, with the probable exception of Eta Carinæ, the most powerful star known to us, outshining the Sun by a factor of at least a million.

The two Magellanic Clouds are about 80,000 light-years apart, and are linked by a stream of gas which also extends in our direction.

Reticulum, adjoining Dorado, is a fairly compact constellation, which makes it easy to identify. Zeta is a very wide double, and though the separation is over 300 seconds of arc the two components have common proper motion, indicating a true association. Beta has an eighth-magnitude companion at a separation of 1,440 seconds of arc; this is an optical pair, not a binary system.

Hydrus (not to be confused with Hydra!) is remarkably devoid of interesting objects. Note that Alpha Hydri is fairly close to Achernar, while Beta is the nearest fairly bright star to the south celestial pole even though it is more than 12 degrees away from the polar point.

Tucana may be the faintest of the Southern Birds, but it is graced by the presence of the Small Magellanic Cloud (SMC), which is further away than the Large Cloud – 190,000 light-years – and is smaller, but is still very prominent with the naked eye. It may well be a double system, with a diameter of at least 16,000 light-years, and like the LMC it contains objects of all kinds; it was by studying short-period variables in it that Henrietta Leavitt was able to establish the law linking the real luminosity of a Cepheid with its period of variation (since for all practical purposes the stars in the SMC may be regarded as equally distant from us). Almost silhouetted against the SMC is the magnificent globular cluster 47 Tucanæ (NGC 104, C106). Of course the alignment is fortuitous, because the cluster belongs to our Galaxy and is in the foreground. On the photograph it looks like a bright star, and indeed the surface brightness is greater than that of the Cloud, as binoculars show; in its way the cluster is more beautiful than Omega Centauri, because it is small enough to fit into a telescopic field.

Octans contains the south celestial pole, but has little else to recommend it; its only star above the fourth magnitude is Nu (the Bayer system of Greek lettering has certainly not been followed here!).

The Mira variable R Octantis is just discernible in the photograph. Not far from it is the Algol eclipsing binary TZ Mensæ, which always remains between magnitudes 6 and 7. Mensa itself is the only one of the 88 accepted constellations which has no star as bright as the fifth magnitude; for the record, its leader is Alpha (5.1, G6). Mensa's sole distinction is that a small part of the LMC extends into it.

DORADO

Brightest star

Star	Proper name	Mag.	Spectrum	RA h m s	Dec. ° ' "
α Alpha		3.27	A0	04 33 59.8	−55 02 42

Variable stars

	RA h m	Dec. ° '	Range (mags.)	Type	Period (d)	Spectrum
R	04 36.8	−62 05	4.8–6.6	Semi-reg.	338	M
β	05 33.6	−62 29	3.7–4.1	Cepheid	9.84	F–G

Galaxy

M/C	NGC	RA h m	Dec. ° '	Mag.	Dimensions min.	Type
	LMC	05 23.6	−69 45	0.1	650 × 550	

Large Cloud of Magellan. Contains 30 Doradûs and 3 planetary nebulæ, NGC 1714, 1722 and 1743

Nebulæ

M/C	NGC	RA h m	Dec. ° '	Dimensions min.	Mag. of illum. star
C103	2070	05 38.7	−69 06	40 × 25	30 Doradûs.

(In the LMC; Tarantula Nebula)

RETICULUM

Brightest stars

Star	Proper name	Mag.	Spectrum	RA h m s	Dec. ° ' "
α Alpha		3.35	G6	04 14 25.5	−62 28 26
β Beta		3.85	K0	03 44 12.0	−64 48 26

Variable star

	RA h m	Dec. ° '	Range (mags.)	Type	Period (d)	Spectrum
R	04 33.5	−63 02	6.5–14.0	Mira	278.3	M

Double stars

	RA h m	Dec. ° '	PA °	Sep. sec	Mags.	
β	03 44.2	−64 48	following	1440	3.9, 8.1	Optical pair
ζ	03 18.2	−62 30	218	310	4.7, 5.2	

TUCANA

Brightest stars

Star	Proper name	Mag.	Spectrum	RA h m s	Dec. ° ' "
α Alpha		2.86	K3	22 18 30.1	−60 15 35
β Beta		3.7	B8+A2	00 32 33	−62 57 43
		(combined)			

Double stars

	RA h m	Dec. ° '	PA °	Sep. sec	Mags.	
β¹	00 31.5	−62 58	169	27.1	4.4, 4.8	
β²	00 31.6	−62 58	295	0.6	4.8, 6.0	Binary, 44.4 y
κ	01 15.8	−68 53	336	5.4	5.1, 7.3	

Globular clusters

M/C	NGC	RA h m	Dec. ° '	Diameter min.	Mag.	
C106	104	00 24.1	−72 05	30.9	4.0	47 Tucanæ
C104	362	01 03.2	−70 51	12.9	6.6	

Galaxy

	RA h m	Dec. ° '	Mag.	Dimensions min.	Type
Small Cloud of Magellan	00 53	−72 50	2.3	280 × 160	

OCTANS

Brightest star

Star	Proper name	Mag.	Spectrum	RA h m s	Dec. ° ' "
ν Nu		3.76	K0	21 41 28.6	−77 23 24

Variable stars

	RA h m	Dec. ° '	Range (mags.)	Type	Period (d)	Spectrum
R	05 26.1	−86 23	6.4–13.2	Mira	405.6	M
U	13 24.5	−84 13	7.1–14.1	Mira	302.6	M
S	18 08.7	−86 48	7.3–14.0	Mira	258.9	M
ε	22 20.0	−80 26	4.9–5.4	Semi-reg.	55	M

Double stars

	RA h m	Dec. ° '	PA °	Sep. sec	Mags.
τ	12 55.0	−85 07	230	0.6	6.0, 6.5
μ²	20 41.7	−75 21	017	17.4	7.1, 7.6
λ	21 50.9	−82 43	070	3.1	5.4, 7.7

MENSA

No star above magnitude 4

Variable star

	RA h m	Dec. ° '	Range (mags.)	Type	Period (d)	Spectrum
TZ	05 30.2	−84 47	6.2–6.9	Algol	8.57	B

HYDRUS

Brightest stars

Star	Proper name	Mag.	Spectrum	RA h m s	Dec. ° ' "
β Beta		2.8	G1	00 25 46.0	−77 15 15
α Alpha		2.86	F0	01 58 46.2	−61 34 12
γ Gamma		3.24	M0	03 47 14.5	−74 14 20

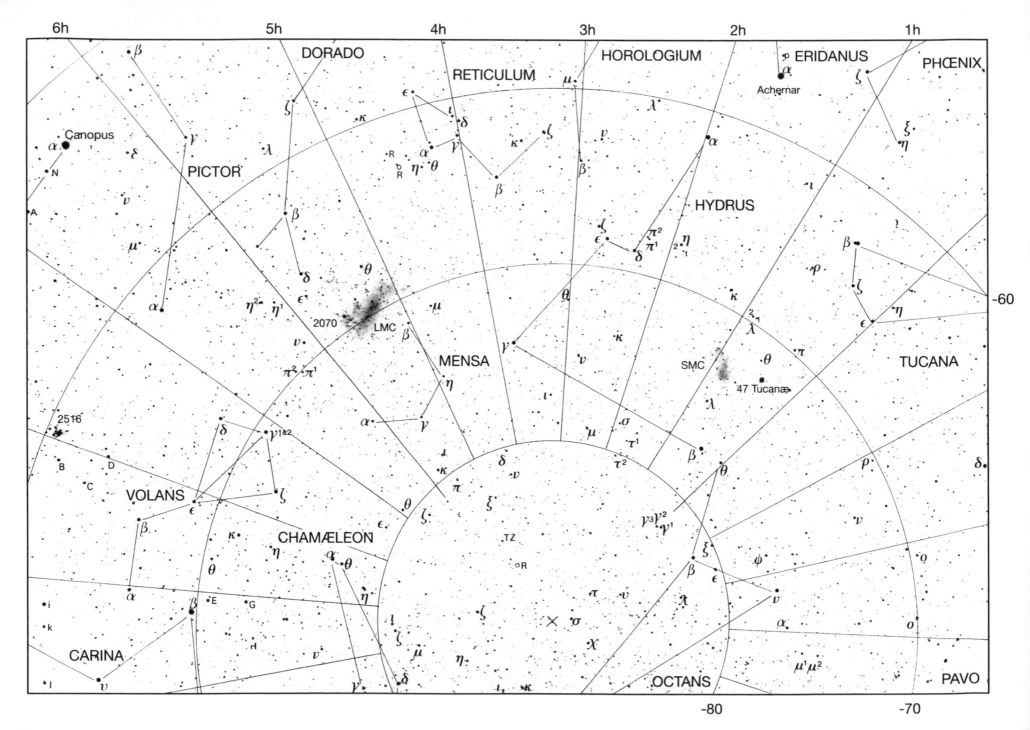

Map 11

CAMELOPARDALIS

Unquestionably this is one of the least interesting areas in the entire sky. Much of it is occupied by Camelopardalis (also called Camelopardus), but the celestial Giraffe has no star above magnitude 4 and no definite pattern; neither are there any striking telescopic objects. Part of Ursa Minor is shown, including Polaris; there are portions of Draco, Lynx, Cepheus, Auriga and Perseus, but that is all. It is worth noting the orange hue of the K-type Lambda Draconis at the top left-hand corner of the photograph; in the lower left-hand corner Beta Aurigæ is shown, just above which is another red star, Pi Aurigæ, of type M. Alpha Persei is also on the photograph, near the bottom right-hand corner.

There are several small-range variables in Camelopardalis. RV is a semi-regular star which fluctuates between about magnitudes 7 and 8; it is of type M, and the colour is pronounced enough when observed through a telescope even though it does not come out well in the photograph simply because the star is not bright enough. BE, which forms a triangle with Beta and Alpha, is clearly orange-red, but it fluctuates only between magnitudes 4.4 and 4.5. BU, which is one of a trio of dim stars, has an equally small range. VZ Camelopardalis, some way from Polaris, has a range of from magnitude 4.8 to 5.2, and is quite prominent in the photograph because of its redness; it is of type M. All that can really be said about Camelopardalis is that some of its leaders – Alpha (4.3, O9.5); Beta (4.3, G8) and 2 or H (4.2, B9) are very remote and luminous; Alpha could match 20,000 Suns.

Much of Ursa Minor is shown, including Polaris, but the whole constellation appears on Map 33, and is described there.

Camelopardalis is not one of Ptolemy's original 48 constellations; it was introduced to the sky by Hevelius of Danzig, in his maps of 1690. It was inevitable that Ptolemy's original list had to be extended, if only because the far-southern part of the sky is not visible from Alexandria, where Ptolemy spent his life; but it must be said that not all of the additions have been happy, and it was Sir John Herschel who commented that the constellation patterns seemed to have been designed so as to cause as much confusion and inconvenience as possible!

Camelopardalis is a case in point; there are no bright stars, and no definite pattern. The other groups introduced by Hevelius were:

Canes Venatici (the Hunting Dogs)
Vulpecula (the Fox; originally Vulpecula et Anser, the Fox and Goose)
Lacerta (the Lizard)
Leo Minor (the Little Lion)
Lynx (the Lynx)
Scutum Sobieskii (Sobieski's Shield)
Monoceros (the Unicorn)
Sextans (the Sextant; originally Sextans Uraniæ, Urania's Sextant)

All of these have survived; some later additions by other astronomers have been rejected. Today, the sky contains 88 officially-accepted constellations, very unequal in size and importance. The largest (Hydra) covers 1,303 square degrees; the smallest (Crux) only 68 square degrees.

CAMELOPARDALIS
No star above magnitude 4

Variable stars

	RA		Dec.		Range	Type	Period	Spectrum
	h	m	°	′	(mags.)		(d)	
RV	04	30.7	+57	25	7.1–8.2	Semi-reg.	101	M
VZ	07	31.1	+82	25	4.8–5.2	Semi-reg.	24	M
R	14	17.8	+83	50	7.0–14.4	Mira	270.2	S

Open cluster

M/C	NGC	RA		Dec.		Diameter	Mag.	No of
		h	m	°	′	min.		stars
	1502	04	07.7	+62	20	8	5.7	45

Galaxies

M/C	NGC	RA		Dec.		Mag.	Dimensions	Type
		h	m	°	′		min.	
	IC 342	03	46.8	+68	06	9.2	17.8 × 17.4	SBc
C7	2403	07	36.9	+65	36	8.4	17.8 × 11.0	Sc

Star trails around the North Celestial Pole

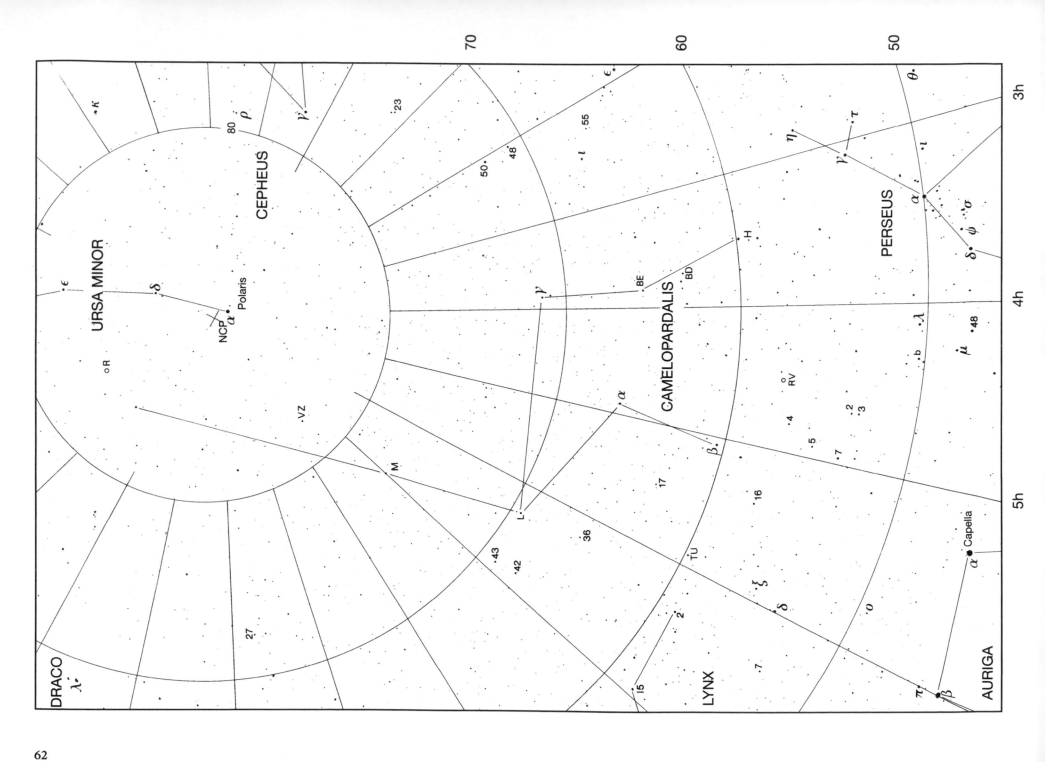

Map 12

PERSEUS

The main constellation in this map is Perseus, which contains a wealth of interesting objects. Auriga is also included, but is described in Map 13; Triangulum is shown, together with parts of Taurus, Aries and Cassiopeia. (Note the orange hue of Gamma Andromedæ, Iota Aurigæ and, of course, Aldebaran, which with the Hyades cluster is at the bottom left-hand corner.) The region is crossed by the Milky Way, and is decidedly rich.

The leading star of Perseus is Alpha (Mirphak), an F-type giant 6,000 times as luminous as the Sun and 620 light-years away. In the photograph it is clearly "off-white", though a yellow cast is not pronounced. However, the most famous star in Perseus is Beta (Algol), the prototype eclipsing binary; apart from Mira Ceti, it was the first star known to vary in light. It has always been known as the Demon Star, because it lies in the head of Medusa, the Gorgon – in mythology a hideous creature with a woman's head and hair of snakes, and whose glance would turn any living thing to stone; Medusa was eventually decapitated by the hero Perseus. However, it does not seem that the old astronomers were aware of its unusual behaviour, and the variations were first noted in 1669 by the Italian astronomer Geminino Montanari; they were explained in 1782 by John Goodricke, who realized that Algol is not intrinsically variable at all. It is made up of two components A and B, so that when the primary is partially hidden by the secondary we witness a drop in light. Normally Algol is of magnitude 2.1, but every 2½ days it starts to fade, dropping to magnitude 3.4 in just over four hours; it remains at minimum for a mere twenty minutes, and takes a further four hours to regain its lost lustre. The

precise interval between successive minima is 2 days 20 hours 48 minutes 56 seconds. The eclipses are only 76 per cent total, so that even at minimum a little light from A still reaches us. When B is hidden by A there is a shallow secondary minimum, but this amounts to no more than a tenth of a magnitude. In the photograph no eclipse is going on, so that Algol is little inferior to Alpha.

We now have a good knowledge of the make-up of the Algol system. A is of spectral type B, about a hundred times as luminous as the Sun and with a diameter of about 3.9 million kilometres (2.4 million miles). The secondary star is of type G, and 3 times as luminous as the Sun; its diameter is over 5 million kilometres (3 million miles), though it is less massive than the primary. The distance between the two is about 10 million kilometres (6 million miles) – too small for them to be seen separately – and the distance from us is 94 light-years. There is spectroscopic evidence of a third member of the system (Algol C) and probably of a fourth.

It is interesting to follow Algol's variations; naked-eye observation is quite adequate, and there are suitable comparison stars in Alpha Persei (1.8), Gamma Andromedæ (2.1), Zeta Persei (2.8) and Beta Trianguli (3.0). Avoid using Rho Persei, which is not far from Algol, but is itself variable – and as would be expected from its M-type spectrum, is very red. Rho is a semi-regular star, with a rough period of around 40 days and an extreme magnitude range of from 3 to 4; useful comparisons, both in Perseus, are Kappa (3.8, K0) and Xi (4.0, O7).

Epsilon, Eta and Theta are wide doubles, but not too easy to resolve because of the faintness of the companions

(little above the tenth magnitude in the case of Theta).

Zeta Persei is very powerful, and could match at least 16,000 Suns; it is well over 1,000 light-years away. It has several faint companions, all below the ninth magnitude, but is of note mainly because it is the brightest member of what is termed a stellar association, made up of a number of hot O and B type stars expanding outwards from a common centre at a rate of over 11 kilometres (7 miles) per second. Presumably the association is young, since it could scarcely have taken more than about a million years to reach its present size at the current rate of expansion. One particularly interesting member of it is X Persei, which can just be made out on the photograph. This is a very "early" type star (spectrum O9.5) and is one of the very few individual stars known to be a source of X-rays. It fluctuates irregularly between magnitudes 6 and 7, and may be compared with the star closely to the right of it in the photograph, which is of magnitude 6.1.

The double cluster NGC 869-884 (C14) is the "show-piece" of Perseus; it is known as the Sword-Handle, or as Chi-h Persei. With the naked eye it appears as a misty patch in the Milky Way, but binoculars show its true nature, and in almost any telescope it makes a glorious spectacle. Both clusters are seen in the photograph and appear slightly bluish; this is not surprising, because the clusters are young, so that their leading stars are hot and white or blue. Each cluster is about 70 light-years in diameter; the distance from us is of the order of 8,000 light-years. The open cluster NGC 957 lies nearby; it pales by comparison, but is easy to see with binoculars.

Another open cluster is M34, not far from Algol;

it too is visible with binoculars, and appears on the photograph as a small patch. The open cluster NGC 1245, between Alpha and Kappa, is rather scattered, but can be identified in the photograph; NGC 1342, between Algol and Zeta, is irregular and sparse, but there should be less trouble in identifying NGC 1528, near Lambda (4.3, B9) in the northern part of the constellation; the adjacent pair of faint stars is a good guide to it.

NGC 1275 (C24), about two degrees east of Algol, is a radio source, and has an appropriate designation (Perseus A); it also emits X-rays. It is the brightest member of a small cluster of galaxies, and it is a very active system, but visually it is elusive, and is barely traceable in the photograph. Also very dim is the California Nebula, NGC 1499 adjoining Xi (4.0, O7); its surface brightness is low, and photography is needed to bring out its complicated structure. It is probably illuminated by Xi, which is a member of the Zeta Persei association.

PERSEUS

Brightest stars

Star		Proper name	Mag.	Spectrum	RA h m s	Dec. ° ' "
α	Alpha	Mirphak	1.80	F5	03 24 19.3	+49 51 40
β	Beta	Algol	2.12 max.	B8	03 08 10.1	+40 57 21
ζ	Zeta	Atik	2.85	B1	03 54 07.8	+31 53 01
ε	Epsilon		2.89	B0.5	03 57 51.0	+40 00 37
γ	Gamma		2.93	G8	03 04 47.7	+53 30 23
δ	Delta		3.01	B5	03 42 55.4	+47 47 15
ρ	Rho	Gorgonea Terti	3.2 max.	M4	03 05 10.5	+38 50 25
η	Eta	Miram	3.76	K3	02 50 41.8	+55 53 44
ν	Nu		3.77	F5	03 45 11.6	+42 34 43
κ	Kappa	Misam	3.80	K0	03 09 29.7	+44 51 27
o	Omicron	Ati	3.83	B1	03 44 19.1	+32 17 18
τ	Tau	Kerb	3.95	G4	02 54 15.4	+52 45 45

Variable stars

	RA h m	Dec. ° '	Range (mags.)	Type	Period (d)	Spectrum
ρ	03 05.2	+38 50	3.0–4.0	Semi-reg.	33–55	M
β	03 08.2	+40 57	2.1–3.4	Algol	2.87	B+G
X	03 55.4	+31 03	6.0–7.0	Irreg.		O9.5 X-ray source
AW	04 47.8	+36 43	7.1–7.8	Cepheid	6.46	F–G

Double stars

	RA h m	Dec. ° '	PA °	Sep. sec	Mags.
ζ	03 54.1	+31 53	AB 208	12.9	2.9, 9.5
			AC 286	32.8	11.3
			AD 195	94.2	9.5
			AE 185	120.3	10.2
ε	03 57.9	+40 01	010	8.8	2.9, 8.1

Open clusters

M/C	NGC	RA h m	Dec. ° '	Diameter min.	Mag.	No of stars
	744	01 58.4	+55 29	11	7.9	20
C14 {	869	02 19.0	+57 09	30	4.3	200 ⌉ Sword
{	884	02 22.4	+57 07	30	4.4	150 ⌡ Handle
	957	02 33.6	+57 32	11	7.6	30
M34	1039	02 42.0	+42 47	35	5.2	60
	1245	03 14.7	+47 15	10	8.4	200
	1342	03 28.4	+37 07	15	8	50
	1444	03 49.4	+52 40	4	6.6	
	1513	04 10.0	+49 31	9	8.4	50
	1528	04 15.4	+51 14	24	6.4	40
	1545	04 20.9	+50 15	8	6.2	20

Planetary nebula

M/C	NGC	RA h m	Dec. ° '	Dimensions sec.	Mag.	Mag. of central star
M76	650–1	01 42.4	+51 34	65 × 290	12.2	17
(Little Dumbbell)						

Nebula

M/C	NGC	RA h m	Dec. ° '	Dimensions min.	Mag. of illum. star
	1499	04 00.7	+36 37	145 × 40	4
(California Nebula)					

Galaxies

M/C	NGC	RA h m	Dec. ° '	Mag.	Dimensions min.	Type
	1023	02 40.4	+39 04	9.5	8.7 × 3.3	E7p
C24	1275	03 19.8	+41 31	11.6	2.6 × 1.9	Pec.
(Perseus A)						

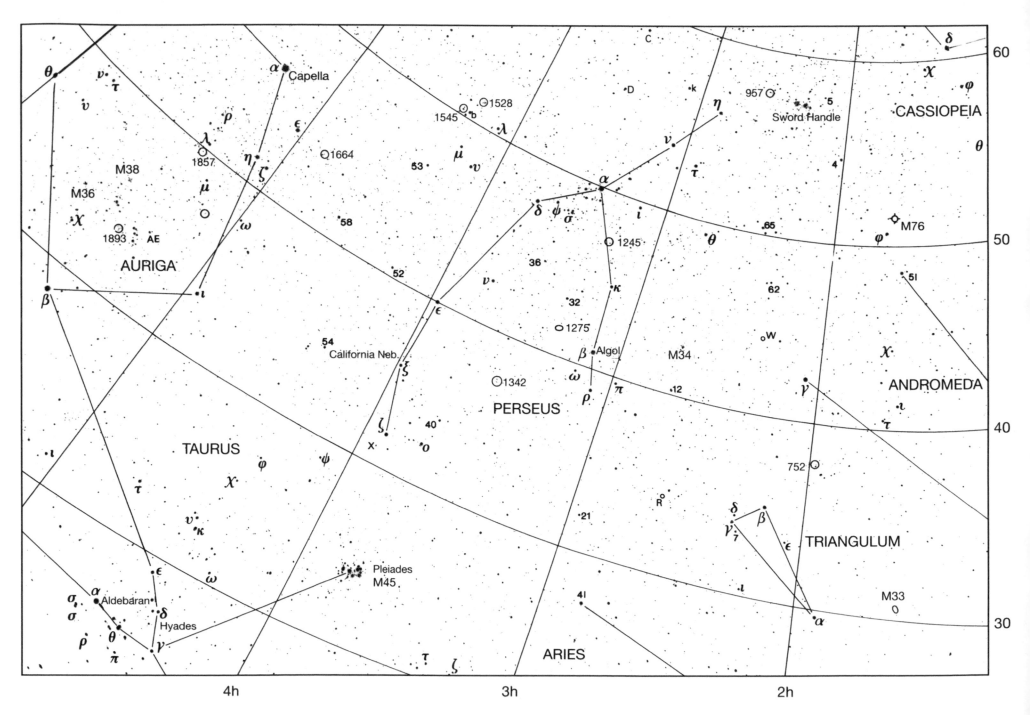

CASSIOPEIA

θ

φ

CASSIOPEIA

θ

Sword Handle

957

5

4

M76

φ

51

ANDROMEDA

χ

ι

τ

TRIANGULUM

δ

β

γ

ε

M33

α

62

W

12

752

R

Capella

α

1528

1545

b

λ

μ

ν

53

η

ν

τ

α

δ ψ σ

ι

θ

1245

36

κ

32

1275

β Algol

M34

ω

ρ π

ζ

X

40

o

21

41

PERSEUS

1342

θ

ν

υ

τ

v

ι

M38

1857

λ

η ζ

μ

ω

58

52

ε

54

California Neb.

ξ

AURIGA

M36

χ

1893

AE

β

ι

TAURUS

ι

τ

φ

ψ

χ

ν

κ

ω

ε

α Aldebaran

σ

σ

δ

Hyades

ρ

θ

γ

π

τ

ζ

ARIES

Pleiades
M45

1664

4h

3h

2h

Map 13
AURIGA

The main constellation here is Auriga, which contains Capella, one of the most brilliant stars in the sky. There are portions of Lynx, Camelopardalis, Perseus, Gemini, Taurus and Orion, all of which are described elsewhere; note the orange colours of Mu and Eta Geminorum and, to a lesser extent, Epsilon. Aldebaran appears to the lower right of the photograph, with the Hyades, and the Crab Nebula (M1), near Zeta Tauri, can just be made out. The red variable U Orionis, between the stars Chi^1 and Chi^2 Orionis, can attain naked-eye visibility, but at the time of this photograph it was near minimum, and does not show up. Not far from Gamma Geminorum, to the lower left of the map, is the red variable BL Orionis. When this picture was taken the star was near maximum (magnitude 6.3); it is also shown in Map 15, but was near its minimum (6.9), and the difference is very obvious.

Auriga is a very prominent constellation, marked by a quadrilateral pattern consisting of Capella, Beta, Theta and Iota; also in the pattern is Al Nath, which was formerly included in Auriga as Gamma Aurigæ, but has now been transferred to Taurus and has become Beta Tauri. Theta is a fairly easy double.

Capella is so far north that it is circumpolar from the latitude of Britain and the northern United States and Canada; it and Vega are on opposite sides of the pole, and at about the same distance from it. Capella is yellow, and is a very close binary; giant telescopes show it as elongated, but in ordinary instruments it looks like a single yellow star. The primary is of type G, while the secondary has an F-type spectrum. Both are large, with diameters of 17.7 million kilometres and 6 million kilometres(11 million and 3.7 million miles) respectively;

the smaller component is only slightly the less massive of the two (2.9 times as massive as the Sun, as against 3.1 solar masses for the primary – remember that smaller stars are always denser than larger ones). The real distance between the two is 110 million kilometres (70 million miles), considerably less than one astronomical unit, and the two components move round their common centre of gravity in a period of 104 days. The distance from us is 43 light-years.

Iota has a K3-type spectrum, and is obviously orange; there are several other orange or reddish stars in the photograph – Pi (4.3, M3) Nu (4.0, K0); Upsilon (4.7, M1) and the rather isolated Delta (3.7, K0).

Close beside Capella lies the triangle of stars made up of Epsilon, Zeta and Eta; the three are known collectively as the Hædi or Kids (in mythology Capella was often called the She-Goat). Eta is a normal B-type star, but the other two are remarkable objects. Epsilon is a particularly luminous F-type supergiant, at least 200,000 times more powerful than the Sun; its distance is over 4,500 light-years, and its usual magnitude is 3.0. However, every 27 years it begins to fade, and takes several months to dim down to magnitude 3.8. It remains at minimum for a year, and then takes months more to recover. Obviously we are dealing with an eclipsing binary rather than an intrinsic variable, but, strangely, the companion has never been traced; it is invisible in ordinary light, shows no spectrum, and is equally coy at other wavelengths. But for its rôle in passing in front of the bright star, we would not know of its existence. It was once thought to be a huge, diffuse star at an early stage of evolution, not yet hot enough to shine, but it is now thought more likely that it

is a smaller, hot star surrounded by an opaque shell of gas and dust which is heated by the inner star and causes the eclipses. The last eclipse ended in 1984, so that there will be no further minima until 2011. Generally, Epsilon is marginally brighter than Eta, though even at maximum it may not be quite constant.

Adjacent to it is another long-period eclipsing binary, Zeta Aurigæ. It is pure coincidence that the two lie side by side, because there is absolutely no connection between them; Zeta is a mere 520 light-years away. Here we have a red supergiant primary, more than 500 times as luminous as the Sun, together with a smaller, hotter companion; the period is 972 days. It is when the hot star is hidden by the supergiant that we see a drop in magnitude, from 3.7 to 4.1; this time the spectrum of the secondary is available for inspection. Before and after the main eclipses, the light from the hot star reaches us after having passed through the very rarefied outer layers of the supergiant, and there are complicated spectral changes which are most informative. When Zeta is faint, a good comparison star is the orange Nu.

Another variable, intrinsic this time, is AE Aurigæ, not far from Iota; it is readily identifiable because of the little chain of stars close beside it. The spectrum is of type O, and the range is small, from magnitude 5.8 to 6.1, so that the star is easy to identify on the photograph. What is not shown, however, is the associated nebulosity, IC 405 (C31). The nebula is illuminated by the star, and is often nicknamed the Flaming Star Nebula; its distance is of the order of 1,600 light-years, and AE itself is a hundred times as luminous as the Sun. Interestingly, it seems that the star is merely passing through the nebula, and does not have a common origin with it. Moreover, AE Aurigæ is one of three so-called "runaway stars" (the others are Mu Columbæ and 53 Arietis) which seem to be moving radially outwards from M42, the Orion Nebula, and seem to have been expelled from it. The Flaming Star Nebula is not easy to see visually, but it shows up well on photographs.

There are three prominent open clusters in Auriga: M36, 37 and 38, all of which are visible with the naked eye and are in fact more prominent than they appear on the photograph. M37, the brightest, is in the same low-power binocular field with Theta, and is easily resolvable with a small telescope. M36 is more condensed, while M38 is larger and looser. All three are between 2,700 and 3,700 light-years away. There are also three less obvious open clusters, NGC 1857, 1778 and 1893, which can be seen on the photograph but are not too easy to identify.

AURIGA

Brightest stars

Star		Proper name	Mag.	Spectrum	RA h m s			Dec. ° ' "		
α	Alpha	Capella	0.08	G8	05	16	41.3	+45	59	53
β	Beta	Menkarlina	1.90	A2	05	59	31.7	+44	56	51
θ	Theta		2.62	A0	05	59	43.2	+37	12	45
ι	Iota	Hassaleh	2.69	K3	04	56	59.5	+33	09	58
ε	Epsilon	Almaaz	2.99v	F0	05	01	58.1	+43	49	24
η	Eta		3.17	B3	05	06	30.8	+41	14	04
δ	Delta		3.72	K0	05	59	31.6	+54	17	05
ζ	Zeta	Sadatoni	3.75v	K4	05	02	28.6	+41	04	33

Variable stars

	RA h m	Dec. ° '	Range (mags.)	Type	Period (d)	Spectrum
RX	05 01.4	+39 58	7.3–8.0	Cepheid	11.62	F–G
ε	05 02.0	+43 49	2.9–3.8	Eclipsing	9892	F
ζ	05 02.5	+41 05	3.7–4.1	Eclipsing	972.1	K+B
R	05 17.3	+53 35	6.7–13.9	Mira	457.5	M
U	05 42.1	+32 02	7.5–15.5	Mira	408.1	M
RT	06 28.6	+30 30	5.0–5.8	Cepheid	3.73	F–G
WW	06 32.5	+38 27	5.8–6.5	Algol	2.53	A+A
UU	06 36.5	+38 27	5.1–6.8	Semi-reg.	234	N

Double stars

	RA h m	Dec. ° '	PA °	Sep. sec	Mags.
ω	04 59.3	+37 53	359	5.4	5.0, 8.0
R	05 17.3	+53 35	339	47.5	var., 8.6
θ	05 59.7	+37 13	AB 313	3.6	2.6, 7.1
			AC 297	50.0	10.6

Open clusters

M/C	NGC	RA h m	Dec. ° '	Diameter min.	Mag.	No of stars
	1664	04 51.1	+43 42	18	7.6	
	1778	05 08.1	+37 03	7	7.7	25
	1857	05 20.2	+39 21	6	7.0	40
	1893	05 22.7	+33 24	11	7.5	60
M38	1912	05 28.7	+35 50	21	6.4	100
M36	1960	05 36.1	+34 08	12	6.0	60
M37	2099	05 52.4	+32 33	24	5.6	150
	2281	06 49.3	+41 04	15	5.4	30

Nebula

M/C	NGC	RA h m	Dec. ° '	Dimensions min.	Mag. of illum. star
	IC 405	05 16.2	+34 16	30 × 19	6v AE Aurigæ

(Flaming Star Nebula)

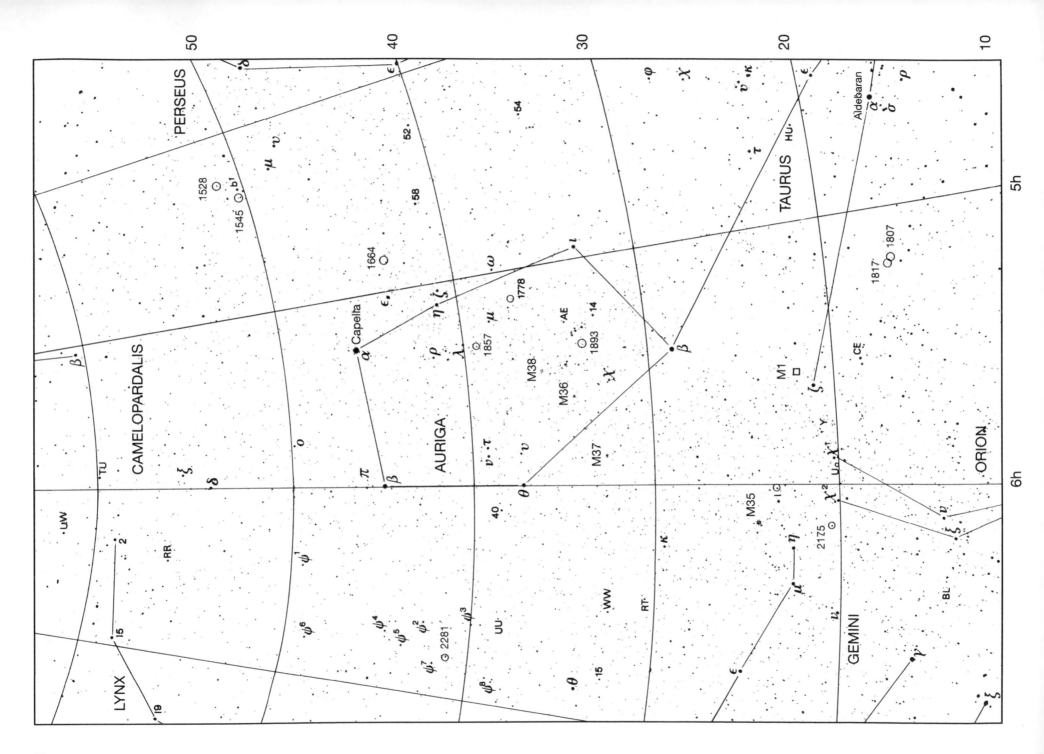

Map 14
TAURUS

The whole of Taurus is shown, together with most of Auriga (see Map 13) and parts of Perseus, Orion, Eridanus and Cetus. Gamma Orionis is seen to the lower left-hand edge of the photograph, and the orange Alpha Ceti at the bottom edge. Note the obvious colour of Iota Aurigæ.

The leader of Taurus is of course Aldebaran, a K-type giant. Outwardly it looks very similar to Betelgeux in Orion, and makes a good comparison for it, though for almost all the time Betelgeux is distinctly the brighter of the two. In fact Aldebaran is not nearly so remote or powerful as Betelgeux; it is 68 light-years away, and only about 90 times as luminous as the Sun, as against 15,000 Sun-power for Betelgeux! Aldebaran gives the impression of being a member of the Hyades cluster, but this is merely a line of sight effect. The cluster lies in the background, with Aldebaran about midway between the Hyades and ourselves.

The V-formation of the Hyades cannot be mistaken (all in all, it is rather a pity that the cluster is so overpowered by the brilliant orange glare of Aldebaran!). The brightest stars in the cluster itself are Theta^2 (3.4, A7); Epsilon (3.5, K0); Gamma (3.6, K0); Theta^1 (3.8, K0); Delta^1 (3.8, K0); and Sigma^2 (4.5, A5). The photograph shows that all the K-type stars are distinctly orange. Delta^1 makes up a wife pair with its neighbour Delta^2 or 64 Tauri (4.8, A7); binoculars show the contrasting colours well. The colour contrast with the two Thetas is even more striking, particularly as this is a naked-eye double; the brighter star is white, the fainter member orange – though binoculars are needed to show the effect well. The white star is the closer to us by 15

light-years, though no doubt all the members of the Hyades have had a common origin. Sigma, close to Aldebaran, consists of two much dimmer stars, both of spectral type A.

Binoculars show many fainter stars in the Hyades (and in fact the cluster is best seen with wide-field binoculars). The total population amounts to several hundreds of stars, but there is no visible nebulosity; the cluster is fairly old, which is why several of the leaders have had time to leave the Main Sequence and become red giants. Star formation has evidently ceased. The situation is very different with the second famous cluster in Taurus, the Pleiades, often nicknamed the Seven Sisters. Quite a number of stars are shown in the photograph; keen-eyed people can see more than seven without optical aid, and the record is said to be held by the nineteenth-century German astronomer Eduard Heis, who could apparently count nineteen. Any optical aid shows many more, and it has been estimated that there are between 300 and 500 stars contained in a spherical area no more than 50 light-years in diameter; if the numbers of faint stars have been under-estimated, even this value may be much too low.

The brightest star of the Pleiades is Eta Tauri or Alcyone (2.9, B7). Then follow 27, Atlas (3.6, B8); 17, Electra (3.7, B6); 20, Maia (3.9, B7); 23, Merope (4.2, B6);19, Taygete (4.3, B6); 28, Pleione (5.1, variable, B8); 16, Celæno (5.4, B7); and 21, Asterope (5.7, B8). Atlas and Pleione lie close together, but can be separated easily enough with binoculars; Pleione is an unstable star which is variable over a small range, and has a variable star designation (BU Tauri).

The fact that all the leading Pleiads are hot white or bluish stars of type B indicates that the cluster is young, and there is a great deal of nebulosity, visually elusive but surprisingly easy to photograph. Star formation is still presumably in progress.

Other open clusters in Taurus, traceable in the photograph, are NGC 1647 and 1746; neither is spectacular, though NGC 1647 contains at least 20 stars. NGC 1817 is fainter and less obvious, best identified from the curved line of stars beside it. NGC 1807 is so sparse and loose that it may be nothing more than an "asterism" – that is to say a chance grouping of unassociated stars rather than a genuine cluster.

Near Zeta Tauri lies M1, the Crab Nebula, known to be the remnant of a supernova which flared up in the year 1054 and was documented by Chinese, Korean and Japanese astronomers; for a time it was visible with the naked eye in broad daylight. The distance from us is 6,000 light-years, so that the actual outburst happened long before there were any terrestrial observers capable of recording it. In the image it is detectable as a very faint patch; it can be seen with powerful binoculars, and almost any small telescope will show it, but photography is needed to bring out the complicated structure which led the Earl of Rosse, in 1850, to give it its familiar nickname. Astronomically it is of supreme importance, because it radiates over almost the whole range of wavelengths, and was one of the first discrete radio sources to be identified.

Nebulosity is also associated with T Tauri, which is the prototype star of its class; it is so young that it is still contracting toward the Main Sequence, and is fluctuating irregularly. Its magnitude range is from 8.4 to 13.5, and on the photograph it is barely traceable, though it was near maximum when the image was taken. The nebulosity is known officially as NGC 1554–5, but more generally as Hind's Variable Nebula, since both nebula and star were discovered in 1852 by the English astronomer John Russell Hind. The star's variations affect the nebula, but even at its best the nebula is a very elusive object, and it is not shown in the photograph.

Kappa Tauri makes up a very wide pair with 67 Tauri; the separation is well over 300 seconds of arc, and, not surprisingly, this is an optical pair rather than a binary. Chi Tauri is an easy double, and Phi has an eighth-magnitude companion.

Lambda Tauri is an Algol-type eclipsing binary, with a range of from magnitude 3.3 to 3.8 and a period of almost 4 days; naked-eye observations are easy, and suitable comparisons are Gamma (3.6), Xi (3.7) and the yellowish Omicron (3.6). Lambda is the brightest of the Algol stars apart from Algol itself. It is 330 light-years away; the primary is a B-type star 450 times as luminous as the Sun, while the A-type secondary has 70 times the Sun's power. As with Algol, the eclipses are partial; at maximum phase about 40 per cent of the primary is obscured. Also as with Algol, the secondary minimum, when the fainter member is partly hidden, is too slight to be detected without photometric equipment. HU Tauri, north of Aldebaran, is also an Algol eclipsing system, identifiable in the photograph; its maximum magnitude is just above 6. CE Tauri, south of Zeta, is a semi-regular variable; it has a small range, but it is of spectral type M, and its colour comes out well in the photograph.

TAURUS

Brightest stars

Star		Proper name	Mag.	Spectrum	RA h m s	Dec. ° ' "
α	Alpha	Aldebaran	0.85v	K5	04 35 55.2	+16 30 33
β	Beta	Al Nath	1.65	B7	05 26 17.5	+28 36 27
η	Eta	Alcyone	2.87	B7	03 47 29.0	+24 06 18 (Pleiades)
ζ	Zeta	Alheka	3.00	B2	05 37 38.6	+21 08 33
λ	Lambda		3.3 max.	B3	04 00 40.7	+12 29 25
θ²	Theta²		3.42	A7	04 28 39.6	+15 52 15 (Hyades)
ε	Epsilon	Ain	3.54	K0	04 28 36.9	+19 10 49 (Hyades)
o	Omicron		3.60	G8	03 24 48.7	+09 01 44
	27	Atlas	3.63	B8	03 49 09.7	+24 03 12 (Pleiades)
γ	Gamma	Hyadum Primus	3.63	K0	04 19 47.5	+15 37 39 (Hyades)
	17	Electra	3.70	B6	03 44 52.5	+24 06 48 (Pleiades)
ξ	Xi		3.74	B8	03 27 10.1	+09 43 58
δ¹	Delta¹		3.76	K0	04 22 56.0	+17 32 33 (Hyades)
θ¹	Theta¹		3.85	K0	04 28 34.4	+15 57 44 (Hyades)
	20	Maia	3.88	B7	03 45 49.5	+24 22 04 (Pleiades)
ν	Nu		3.91	A1	04 03 09.3	+05 59 21

Variable stars

	RA h m	Dec. ° '	Range (mags.)	Type	Period (d)	Spectrum
BU	03 49.2	+24 08	4.8–5.5	Irreg.		Bp Pleione
λ	04 00.7	+12 29	3.3–3.8	Algol	3.95	B+A
T	04 22.0	+19 32	8.4–13.5	T Tauri	Irreg.	G–K
R	04 28.3	+10 10	7.6–14.7	Mira	323.7	M
HU	04 38.3	+20 41	5.9–6.7	Algol	2.06	A
TU	05 45.2	+24 25	5.9–8.6	Semi-reg.	190	N
SU	05 49.1	+19 04	9.1–16.0	R Coronæ		G0p

Double stars

	RA h m	Dec. ° '	PA °	Sep. sec	Mags.	
χ	04 22.6	+25 38	024	19.4	5.5, 7.6	
66	04 23.9	+09 28	265	0.1	5.8, 5.9	Binary, 516 y
κ+67	04 25.4	+22 18	173	339	4.2, 5.3	
θ	04 28.7	+15 32	346	337.4	3.4, 3.8	
σ	04 39.3	+15 55	193	431.2	4.7, 5.1	
126	05 41.3	+16 32	238	0.3	5.3, 5.9	

Open clusters

M/C	NGC	RA h m	Dec. ° '	Diameter min.	Mag.	No of stars
M45	1432/5	03 47.0	+24 07	110	1.2	300 + (Pleiades)
C41		04 27.0	+16	330	1.0	200 + (Hyades)
	1647	04 46.0	+19 04	45	6.4	200
	1746	05 03.6	+23 49	42	6.1	20
	1807	05 10.7	+16 32	17	7.0	20 Asterism?
	1817	05 12.1	+16 42	16	7.7	60

Nebulæ

M/C	NGC	RA h m	Dec. ° '	Dimensions min.	Mag. of illum. star
	1554–5	04 21.8	+19 32	var.	9v
(Hind's Variable Nebula associated with T Tauri)					
M1	1952	05 34.5	+22 01	6 × 4	16
(Crab Nebula: SNR)					

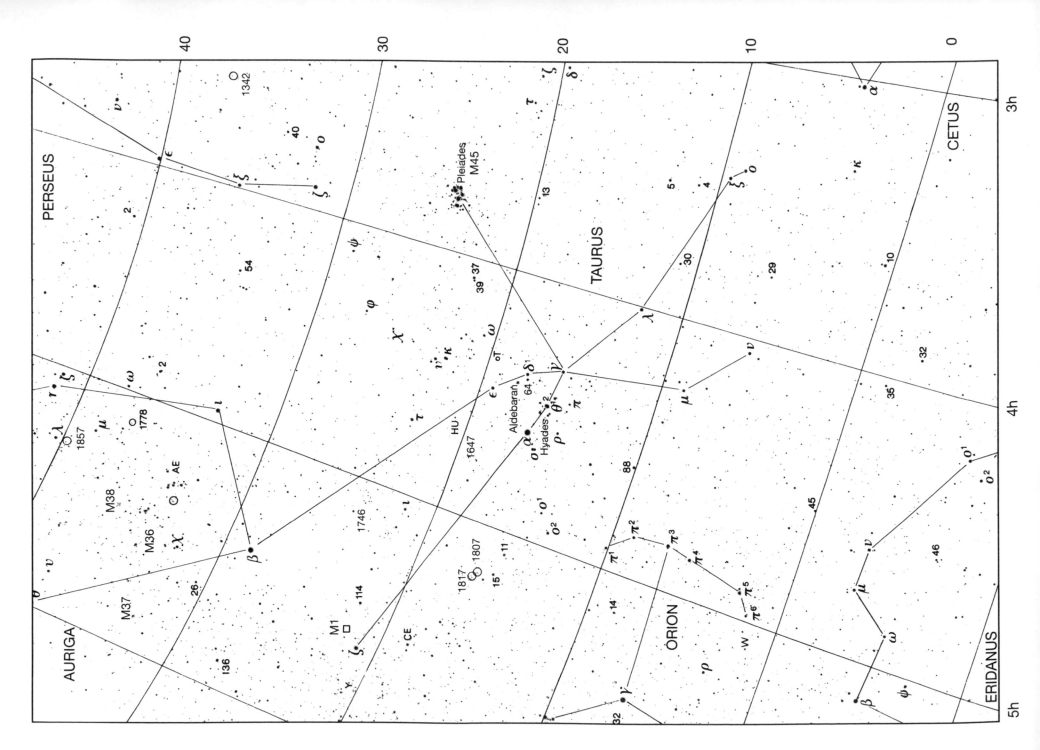

74

Map 15

ORION

With this map we come to Orion, which is often – and perhaps rightly – regarded as the most splendid of all the constellations. With its characteristic shape, its starry Belt, its misty Sword, and its two brilliant leaders Betelgeux and Rigel it cannot possibly be mistaken, and since it is crossed by the celestial equator it is visible from any part of the world. The map also includes most of Monoceros (not all), much of Taurus, and parts of Gemini and Eridanus. Note the orange Nu and Eta Geminorum, near the upper left edge of the photograph; Zeta Tauri, with the faint impression of the Crab Nebula; and of course Aldebaran, with the Hyades.

Though lettered Beta, Rigel is the brightest star in Orion, and is usually several tenths of a magnitude superior to Alpha (Betelgeux). Rigel is a true cosmical searchlight, at least 60,000 times more luminous than the Sun, and around 900 light-years away. It is of spectral type B8, and is bluish-white, with a very hot surface. It is squandering its fuel reserves at a furious rate, and in its present form can hardly last for more than a few million years; since its mass is of the order of 25 times that of the Sun, it may eventually explode as a supernova. It has a 6.8-magnitude companion at a separation of over 9 seconds of arc; it is an easy telescopic object, and is itself a close binary. It is genuinely associated with Rigel, though the two are a very long way apart – perhaps over 3,000 astronomical units.

Alpha Orionis (Betelgeux) is quite different. It is a huge red supergiant, and though it is not as powerful as Rigel it could still match 15,000 Suns. Its diameter is of the order of 400 million kilometres (250 million miles), so that it could comfortably contain the entire

orbit of the Earth round the Sun. It is variable; the general magnitude range is between 0.4 and 0.8, but there have been occasions when it has almost equalled Rigel, as indeed it did during the latter part of 1995. The variations are intrinsic; Betelgeux swells and shrinks, changing its output of energy as it does so. It is classed as semi-regular, with a period of just over 2,100 days, but there are very marked irregularities, and the star is never predictable. When the photograph was taken it was not far from maximum, so that it was much brighter than Aldebaran and not greatly inferior to Rigel.

All the other leaders of Orion are hot and bluish-white; Kappa, in particular, is not much less luminous than Rigel, but is further away (over 2,000 light-years). Delta, the northernmost star of the Belt, lies very close to the celestial equator, and is an eclipsing binary with a very small range (2.20 to 2.35); it has a visual companion of just below the sixth magnitude, and since the separation is over 50 seconds of arc this is an easy pair to resolve. The secondary is also of type B. It seems to be at least half a light-year from the primary, but since the two stars share a common motion in space it is likely that the association is genuine rather than a line of sight effect.

It is interesting to compare Epsilon and Zeta, in the Belt. In the photograph Epsilon is the brighter of the two, and the difference is quite obvious with the naked eye, even though it amounts to less than a tenth of a magnitude (1.70 for Epsilon, 1.79 for Zeta). Zeta is double; the companion is of the fourth magnitude. This is a binary system (period 1,509 years) and there is also a more distant tenth-magnitude companion which may or may not be a genuine member of the Zeta system.

Eta Orionis is another double, though the separation is only 1.5 seconds of arc and the pair is hard to resolve with small telescopes. Lambda is a much easier double, and makes up a neat little group with Phi^1 and Phi^2; Phi^2 is of type K, and decidedly orange. Pi^3 – one of a curved line of stars outside the main pattern, all of which are lettered Pi – has an 8.7-magnitude companion at a separation of over 90 seconds of arc.

W Orionis, not far from Pi^6, is a semi-regular variable with a small range; when the photograph was taken it was near its minimum of magnitude 7.7, but its redness is obvious (the spectrum is of type N). CK Orionis, near Gamma, is also a semi-regular, though of type K; it too is clearly coloured. BL Orionis, not far from Gamma Geminorum, is well seen here; compare its brightness in this photograph with the representation in the photograph with Map 13.

U Orionis was at minimum in the Map 13 photograph. Here it is much brighter, and is easy to locate between Chi^1 (4.4, G0) and Chi^2 (4.6, B2); it is an M-type Mira star, with a range of from magnitude 4.8 to 12.6, so that at maximum it may reach naked-eye visibility. The period is 372 days, and so the star reaches its greatest brightness about one week later in every year.

Orion contains the most famous of all the bright nebulæ – M42, in the Hunter's Sword. It is a stellar birthplace, inside which fresh stars are being produced. The gas content is largely hydrogen, and this is why the nebula appears red in the photograph, though visually the colour is not evident; the nebula looks greenish white, with dark patches here and there. M42 is illuminated by the multiple star Theta Orionis which lies on the nearside edge of the nebula; Theta is nicknamed the Trapezium, because of the arrangement of its four principal components. A small telescope shows the Trapezium well, and it is an ideal target for photographers. The nebula is over 1,000 light-years away, and very massive even though its material is almost incredibly tenuous – millions of times less dense than the air we breathe. M43 is really an extension of M42, and slightly beyond to the south is the open cluster NGC 1981, which is not hard to resolve into stars. Also in the Sword, though clear of the main nebulosity, is the multiple Sigma Orionis, and north of the Belt lies the gaseous nebula M78, whose shape has been compared with that of a comet; it is not bright, and is rather hard to identify in the photograph. In fact almost the whole of Orion is contained in a vast molecular cloud, and the bright nebulæ represent only the illuminated parts of it.

The northernmost part of Eridanus, the celestial River, adjoins Orion; Beta Eridani (Kursa), of magnitude 2.8, is not far from Rigel. it is worth seeking out the little pair made up of Omicron[1] (4.0, F2) and Omicron[2] (6.0, K1). Omicron[2] is a triple system; the main pair has a separation of 83 seconds of arc, and makes up a binary combination.

Monoceros, the Unicorn, lies on the opposite side of Orion, occupying most of the large triangle formed by Betelgeux in Orion, Procyon in Canis Minor and Sirius in Canis Major. Monoceros is also shown in Map 18, and is described there.

ORION

Brightest stars

Star		Proper name	Mag.	Spectrum	RA h m s	Dec. ° ' "
β	Beta	Rigel	0.12v	B8	05 14 32.2	−08 12 06
α	Alpha	Betelgeux	0.5v	M2	05 55 10.2	+07 24 26
γ	Gamma	Bellatrix	1.64	B2	05 25 07.8	+06 20 59
ε	Epsilon	Alnilam	1.70	B0	05 36 12.7	−01 12 07
ζ	Zeta	Alnitak	1.77	09.5	05 40 45.5	−01 56 34
κ	Kappa	Saiph	2.06	B0.5	05 47 45.3	−09 40 11
δ	Delta	Mintaka	2.23v	09.5	05 32 00.3	−00 17 57
ι	Iota	Hatysa	2.76	09	05 35 25.9	−05 54 36
π³	Pi³		3.19	F6	04 49 50.3	+06 57 41
η	Eta	Algjebbah	3.36	B1	05 24 28.6	−02 23 50
λ	Lambda	Heka	3.39	08	05 35 08.2	+09 56 02
τ	Tau		3.60	B5	05 17 36.3	−06 50 40
π⁴	Pi⁴		3.69	B2	04 51 12.3	+05 36 18
π⁵	Pi⁵		3.72v	B2	04 54 15.0	+02 26 26
σ	Sigma		3.73	09.5	05 38 44.7	−02 36 00

Variable stars

	RA h m	Dec. ° '	Range (mags.)	Type	Period (d)	Spectrum
W	05 05.4	+01 11	5.9–7.7	Semi-reg.	212	N
S	05 29.0	−04 42	7.5–13.5	Mira	419.2	M
CK	05 30.3	+04 12	5.9–7.1	Semi-reg.	120	K
α	05 55.2	+07 24	0.1–0.9	Semi-reg.	2110	M
U	05 55.8	+20 10	4.8–12.6	Mira	372.4	M
BL	06 25.5	+14 43	6.3–7.0	Irreg.	-	N

Double stars

	RA h m	Dec. ° '		PA °	Sep. sec	Mags.	
β	05 14.5	−08 12		202	9.5	0.1, 6.8	
ρ	05 15.3	+02 54		064	7.0	4.5, 8.3	
η	05 24.5	−02 24	AB 080		1.5	3.8, 4.8	
			AC 051		115.1	9.4	
δ	05 32.0	−00 18		359	52.6	2.2v, 6.3	
λ	05 35.1	+09 56		043	4.4	3.6, 5.5	
σ	05 38.7	−02 36	AB 137		0.2	4.0, 6.0	Binary, 170 y
			AB+C 238		11.4	10.3	
			AB+D 084		12.9	7.5	
			AB+E 061		42.6	6.5	
θ	05 35.3	−05 23	AB 031		8.8	6.7, 7.9	
			AC 132		12.8	5.1	
			AD 096		21.5	6.7	
ι	05 35.4	−05 55		141	11.3	2.8, 6.9	
ζ	05 40.8	−01 57	AB 162		2.4	1.9, 4.0	Binary, 1509 y
			AC 010		57.6	9.9	

Open clusters

M/C	NGC	RA h m	Dec. ° '	Diameter min.	Mag.	No of stars
	1981	05 35.2	−04 26	25	4.6	20
	2175	06 09.8	+20 19	18	6.8	60

Nebulæ

M/C	NGC	RA h m	Dec. ° '	Dimensions min.	Mag. of illum. star
M42 (Great Nebula)	1976	05 35.4	−05 27	66 × 60	5
M43	1982	05 35.6	−05 16	20 × 15	7 Extension of M42
M78	2068	05 46.7	+00 03	8 × 6	10 Nebula is mag. 8
	IC 434	05 41.0	−02 24	60 × 10	2 (ζ)

(Behind Horse's Head dark nebula. Barnard 33)

ERIDANUS

Double star

	RA h m	Dec. ° '	PA °	Sep. sec	Mags.
o²	04 15.2	−07 39	107	83.4	4.4, 9.5

For brightest stars see Map 8.

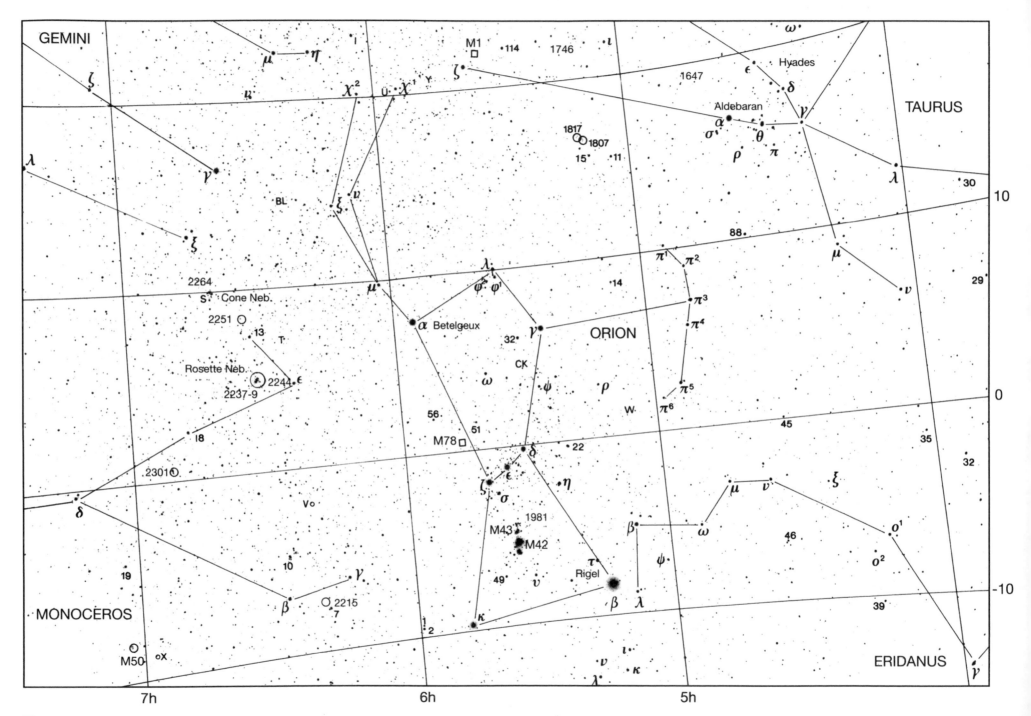

GEMINI

TAURUS

ORION

MONOCEROS

ERIDANUS

Hyades

Aldebaran

Betelgeux

Rigel

Rosette Neb.

Cone Neb.

M1

M50

M78

M43

M42

1817
1807

1981

2264

2251

2244

2237-9

2301

2215

M1

1746

1647

88

114

7h

6h

5h

Map 16

LEPUS, COLUMBA, CÆLUM

This map is of course dominated by Canis Major, led by Sirius, by far the most brilliant star in the entire sky. However, Canis Major is also shown on Map 19, and is described there.

Lepus contains an intensely red variable, R, which has an N-type spectrum and is nicknamed the Crimson Star. At maximum it can rise to magnitude 5.5, but unfortunately it was near minimum when this photograph was taken, so that it does not appear; it lies fairly near Mu (3.3, B9). The semi-regular red variable S Leporis, making a triangle with Delta and Gamma, was, however, at its best (magnitude 7.1) and shows up well. Gamma itself is a wide, easy double; the components are of types F and K respectively, so that there is good colour contrast. Not far from the reddish Epsilon is a fine globular cluster, M79, which is small but bright and condensed. NGC 2017, near the F-type, rather yellowish Alpha, is a large but scattered open cluster.

Columba is not a very prominent constellation, and from the latitudes of northern Europe and the northern United States and Canada it is always low (from Britain, for example, Eta Columbæ does not rise at all). Alpha and Gamma have faint companions; Pi² is double with equal components, but the separation is so small that the pair is far from easy to resolve. The Mira variable T Columbæ is well shown, since at the time of the photograph it was near its maximum brightness (magnitude 6.6). NGC 1851 (C73) is a globular cluster on the borders of Columba and Cælum; it has an integrated magnitude of just below 7, and is barely visible in the photograph. It is notable as being a source of X-rays. Cælum itself is entirely unremarkable; the red semi-regular variable

R Cæli can rise to magnitude 6.7, but was near minimum at the time of the photograph, and is barely visible.

Part of Puppis is also shown here; note the orange hue of Pi. The open cluster M93 lies at the extreme left-hand edge of the map, and NGC 2477 (C71) can also be seen. Puppis is described with Map 20.

A small section of Eridanus is shown here. Eridanus, the celestial River, is an immensely long constellation; of its leading stars, the southernmost (Achernar) lies at declination −57°, the northernmost (Beta) at only −05°. Eridanus is also one of only nine constellations covering more than 1,000 square degrees of the sky. The list of these very large constellations is:

Constellation	Area (square degrees)
Centaurus	1,060
Cetus	1,232
Draco	1,083
Eridanus	1,138
Hercules	1,225
Hydra	1,303
Pegasus	1,121
Ursa Major	1,280
Virgo	1,294

LEPUS

Brightest stars

Star	Proper name	Mag.	Spectrum	RA h m s	Dec. ° ' "
α Alpha	Arneb	2.58	F0	05 32 43.7	−17 49 20
β Beta	Nihal	2.84	G2	05 28 14.7	−20 45 35
ε Epsilon		3.19	K5	05 05 27.6	−22 22 16
μ Mu		3.31	B9	05 12 55.8	−16 12 20
ζ Zeta		3.55	A3	05 46 57.2	−14 49 20
γ Gamma		3.60	F6	05 44 27.7	−22 26 55
η Eta		3.71	F0	05 56 24.2	−14 10 04
δ Delta		3.81	G8	05 51 19.2	−20 52 45

Variable stars

	RA h m	Dec. ° '	Range (mags.)	Type	Period (d)	Spectrum
R	04 59.6	−14 48	5.5–11.7	Mira	432.1	N
RX	05 11.4	−11 51	5.0–7.0	Irreg.		M
S	06 05.8	−24 12	7.1–8.9	Semi-reg.	90	M

Double stars

	RA h m	Dec. ° '	PA °	Sep. sec	Mags.
κ	05 13.2	−12 56	358	2.6	4.5, 7.4
β	05 28.2	−20 46	AB 330	2.5	2.8, 7.3
			AC 145	64.3	11.8
			AD 075	206.4	10.3
			AE 058	241.5	10.3
γ	05 44.5	−22 27	350	96.3	3.7, 6.3

Globular cluster

M/C	NGC	RA h m	Dec. ° '	Diameter min.	Mag.
M79	1904	05 24.5	−24 33	8.7	8.0

COLUMBA

Brightest stars

Star	Proper name	Mag.	Spectrum	RA h m s	Dec. ° ' "
α Alpha	Phakt	2.64	B8	05 39 38.9	−34 04 27
β Beta	Wazn	3.12	K2	05 50 57.5	−35 46 06
δ Delta		3.85	gG1	06 22 06.7	−33 26 11
ε Epsilon		3.87	K0	05 31 12.7	−35 28 15
η Eta		3.96	K0	05 59 08.8	−42 48 55

Variable stars

	RA h m	Dec. ° '	Range (mags.)	Type	Period (d)	Spectrum
T	05 19.3	−33 42	6.6–12.7	Mira	225.9	M
R	05 50.5	−29 12	7.8–15.0	Mira	327.6	M

Double stars

	RA h m	Dec. ° '	PA °	Sep. sec	Mags.
α	05 39.6	−34 04	359	13.5	2.6, 12.3
γ	05 57.5	−35 17	110	33.8	4.4, 12.7
π²	06 07.9	−42 09	150	0.1	6.2, 6.3

Globular cluster

M/C	NGC	RA h m	Dec. ° '	Diameter min.	Mag.	
C73	1851	05 14.1	−40 03	11.0	7.3	X-ray source

CÆLUM

No star above magnitude 4

Variable stars

	RA h m	Dec. ° '	Range (mags.)	Type	Period (d)	Spectrum
R	04 40.5	−38 14	6.7–13.7	Mira	390.9	M
T	04 47.3	−36 13	7.0–9.8	Semi-reg.	156	N

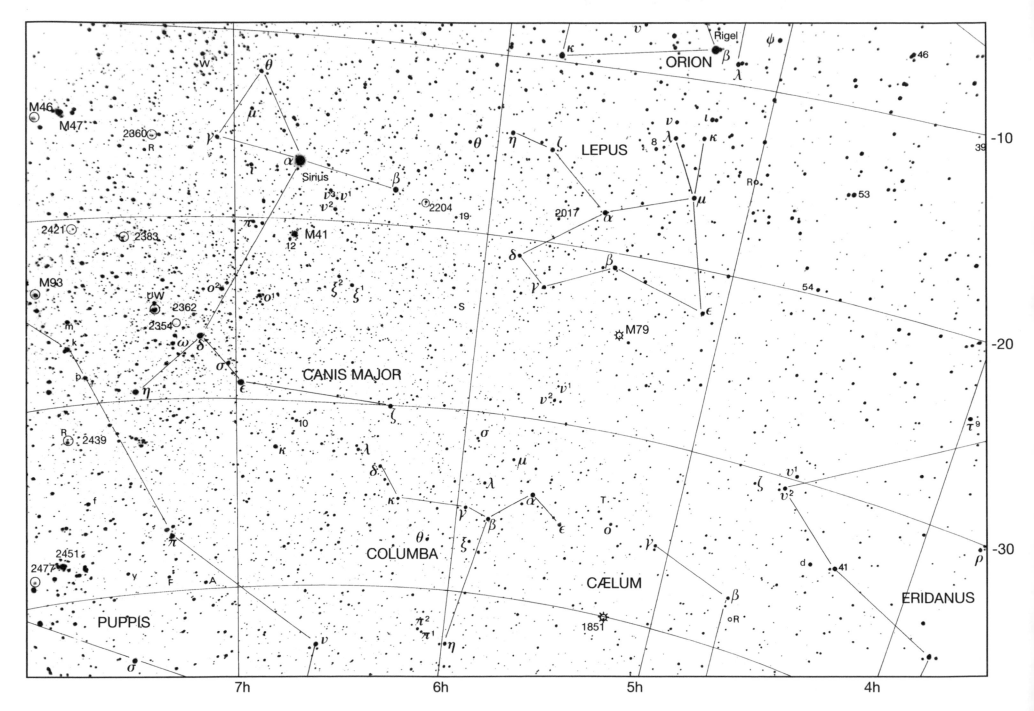

ORION

Rigel

LEPUS

CANIS MAJOR

COLUMBA

CÆLUM

ERIDANUS

PUPPIS

Sirius

M46
M47
2360
R
2421
2383
M93
UW
2362
2354
m
k
p
2439
R
f
2451
2477
y
F
A
σ

W
θ
μ
γ
ι
α
β
ν³ ν¹
ν²
2204
19
M41
12
o²
o¹
ξ² ξ¹
ω
δ
σ
ε
η
ζ
10
κ
λ
δ
κ
γ
β
θ
ξ
π
ν
π²
π¹
η
1851
oR
β
γ
α
ε
ο
τ
λ
μ
σ
ν² ν¹
M79
θ
η
ζ
α
2017
δ
β
γ
ε
ν
8
λ
ι
κ
μ
Ro
53
54
ζ
ν¹
ν²
d
41
ρ
τ⁹
46
39
κ
ν
ψ
λ
S

-10

-20

-30

7h   6h   5h   4h

Map 17

GEMINI, CANIS MINOR

This map contains the whole of Gemini and Canis Minor, together with parts of other constellations – Auriga, Cancer, Orion, Monoceros and Lynx. Præsepe (M44) in Cancer, the celebrated open cluster, is shown to the lower left-hand edge of the photograph, and much of Auriga is included, with the star-clusters M36, M37 and M38; Capella is just off the top of the chart, above the centre. The Crab Nebula (M1) in Taurus is just visible, above Zeta Tauri. A section of Orion also appears; this includes the Mira variable U Orionis, but when this photograph was taken U was near its minimum, and does not appear (compare the view with those on Maps 13 and 15).

Gemini is one of the most brilliant of the Zodiacal constellations. Its main stars are, of course, the Twins – Castor and Pollux – but they are very dissimilar, and are not genuinely associated with each other; Castor is 46 light-years away, Pollux only 36. Neither are they of the same colour, and the whiteness of Castor contrasts with the warm orange glow of Pollux. Today, Pollux (magnitude 1.14) is always included in the official list of "first-magnitude" stars, but Castor (1.58) is not, though the division is quite arbitrary and seems indeed rather illogical. Moreover, Ptolemy and other sky-watchers of Classical times ranked Castor and Pollux equal, of the second magnitude, and in 1700 Flamsteed actually made Castor the brighter of the two. Yet, as is always the case with alleged cases of permanent change, the evidence is very slender indeed.

Pollux is a single K-type star, 60 times as luminous as the Sun, but Castor is a multiple system. A small telescope shows that it is a fine double; the components are of magnitudes 1.9 and 2.9, and the pair makes up a binary pair, with an orbital period of 420 years. The present separation is 2.5 seconds of arc, but this is slowly increasing, and will exceed 4 seconds of arc by the end of the century, so that Castor is a very easy telescopic pair; the real separation between the components is around 90 astronomical units. Both components are spectroscopic binaries, and all four stars are of spectral Type A. There is also a fainter member of the system – Castor C, alternatively known by its variable star designation of YY Geminorum. It too is a spectroscopic binary; both the components are feeble red dwarfs. The combined magnitude is 9, and the separation is over 70 seconds of arc, but YY is not an easy object because of the glare from Castor itself. The actual separation between the red dwarfs is less than 3.2 million kilometres (2 million miles).

Below Pollux on the photograph, and rather to the right, is Kappa Geminorum (3.6, G8). Two more fairly conspicuous stars lie between Kappa and Gamma or Alhena (1.9, A0), in the "foot" of the Twins. Delta Geminorum is a normal F-type star of magnitude 3.5; Zeta is a Cepheid variable, with a range of from magnitude 3.7 to 4.1 and a period of 10.15 days. When this photograph was taken Zeta was near its maximum brightness, and is almost equal to Delta; when it is near minimum a suitable comparison star is Upsilon (4.1, M0), which makes up a small triangle with Pollux and Kappa, and is distinguished by its redness.

Another naked-eye variable is Eta Geminorum, sometimes still called by its old proper name of Propus. It is of type M, and is strongly orange in hue; the range is from magnitude 3.2 to 3.9, and there is a very rough period of about 233 days. A good comparison star is Mu (2.9, M3); as can be seen from the photograph Eta is the more obviously coloured of the two, though they are of virtually the same spectral type. When Eta is faint, it may be compared with 1 Geminorum (4.2, G5) and Nu (4.1, B7). Incidentally, it was close to 1 Geminorum that William Herschel discovered Uranus, in 1781 – admittedly without realizing that he had found anything more exciting than a comet! And in 1930 Clyde Tombagh located Pluto close to Delta, so that all in all Gemini seems to have been a very fruitful region for planet-hunters.

Several telescopic variables are shown in the photograph. R Geminorum, a Mira star, lies inside the little triangle made up of Delta, Zeta and Omega (52, G2); the period is 370 days, and the spectrum is of type S. The red colour is very pronounced, though it does not show up on the photograph because the star is not bright enough. Another Cepheid, W Geminorum, lies near Gamma; it never becomes much fainter than magnitude 7.5. The red irregular variable BU Geminorum can just be seen, near Eta; at the time of the photograph its magnitude also was about 7. Apart from Castor, there are several double stars in Gemini within the range of modest telescopes; Mu, Kappa, Nu and Lambda all have companions which are above the tenth magnitude. Eta has a companion of magnitude 8.8, and Zeta has two faint attendants which are not, however, genuinely associated with it.

Near Eta and 1 is a spectacular open cluster, M35, which is visible with the naked eye. On the photograph it looks decidedly bluish, though when viewed through a telescope the leading stars appear white. Binoculars give a good view of the cluster; the real diameter is about 30 light-years, and it is over 2,000 light-years away. NGC 2129 is a much fainter loose cluster, just traceable on the photograph close to 1 Geminorum.

Another interesting object in Gemini is NGC 2392 (C39), a planetary nebula which lies not far from Delta; it is often nicknamed the Eskimo Nebula, but has also been referred to as the Clown Face. It can just be identified on the photograph, but it is a decidedly elusive telescopic object. The central star, of magnitude 10.5, is exceptionally hot, with a surface temperature of at least 40,000 degrees.

Canis Minor has only one bright star, Procyon. At less than 12 light-years, it is one of our nearest stellar neighbours. Like Sirius, it has a white dwarf companion, but the brilliance of the primary makes the white dwarf a difficult object. Beta (2.9, B8) makes up a neat little triangle with Gamma (4.3, K3) and Epsilon (5.0, G8); in this group is the Mira-type variable S Canis Minoris, barely visible on the photograph, though at its best it can rise to magnitude 6.6.

GEMINI
Brightest stars

Star		Proper name	Mag.	Spectrum	RA h m s	Dec. ° ' "
β	Beta	Pollux	1.14	K0	07 45 18.9	+28 01 34
α	Alpha	Castor	1.58	A0	07 34 35.9	+31 53 18
γ	Gamma	Alhena	1.93	A0	06 37 42.7	+16 23 57
μ	Mu	Tejat	2.88v	M3	06 22 57.6	+22 30 49
ε	Epsilon	Mebsuta	2.98	G8	06 43 55.9	+25 07 52
η	Eta	Propus	3.2 max.	M3	06 14 52.6	+22 30 24
ξ	Xi	Alzirr	3.36	F5	06 45 17.3	+12 53 44
δ	Delta	Wasat	3.53	F2	07 20 07.3	+21 58 56
κ	Kappa		3.57	G8	07 44 26.8	+24 23 52
λ	Lambda		3.58	A3	07 18 05.5	+16 32 25
θ	Theta		3.60	A3	06 52 47.3	+33 57 40
ζ	Zeta	Mekbuda	3.7 max.	G0	07 04 06.5	+20 34 13
ι	Iota		3.79	K0	07 25 43.5	+27 47 53

Variable stars

	RA h m	Dec. ° '	Range (mags.)	Type	Period (d)	Spectrum
BU	06 12.3	+22 54	5.7–7.5	Irreg.		M
η	06 14.9	+22 30	3.2–3.9	Semi-reg.	233	M
W	06 35.0	+15 20	6.5–7.4	Cepheid	7.91	F–G
X	06 47.1	+30 17	7.5–13.6	Mira	263.7	M
ζ	07 04.1	+20 34	3.7–4.1	Cepheid	10.15	F–G
R	07 07.4	+22 42	6.0–14.0	Mira	369.8	S
U	07 55.1	+22 00	8.2–14.9	SS Cygni	±103	M+WD

Double stars

	RA h m	Dec. ° '	PA °	Sep. sec	Mags.	
38	06 54.6	+13 11	147	7.0	4.7, 7.7	Binary, 3190 y
ζ	07 04.1	+20 34	AB 084	87.0	var., 10.5	
			AC 350	96.5	8.0	
δ	07 20.1	+21 59	223	6.0	3.5, 8.2	Binary, 1200 y
α	07 34.6	+31 53	AB 088	2.5	1.9, 2.9	Binary, 420 y
			AC 164	72.5	1.6, 8.8	
κ	07 44.4	+24 24	240	7.1	3.6, 8.1	

Open clusters

M/C	NGC	RA h m	Dec. ° '	Diameter min.	Mag.	No of stars
	2129	06 01.0	+23 18	7	6.7	40
	2169	06 08.4	+13 57	7	5.9	30
M35	2168	06 08.9	+24 20	28	5.0	200
	2395	07 27.1	+13 35	12	8.0	30 Asterism?

Planetary nebula

M/C	NGC	RA h m	Dec. ° '	Dimensions sec.	Mag.	Mag. of central star
C39 (Eskimo Nebula)	2392	07 29.2	+20 55	13 × 44	10	10.5

CANIS MINOR
Brightest stars

Star		Proper name	Mag.	Spectrum	RA h m s	Dec. ° ' "
α	Alpha	Procyon	0.38	F5	07 39 18.1	+05 13 30
β	Beta	Gomeisa	2.90	B8	07 27 09.0	+08 17 21

Variable stars

	RA h m	Dec. ° '	Range (mags.)	Type	Period (d)	Spectrum
R	07 08.7	+10 01	7.3–11.6	Mira	337.8	S
S	07 32.7	+08 19	6.6–13.2	Mira	332.9	M

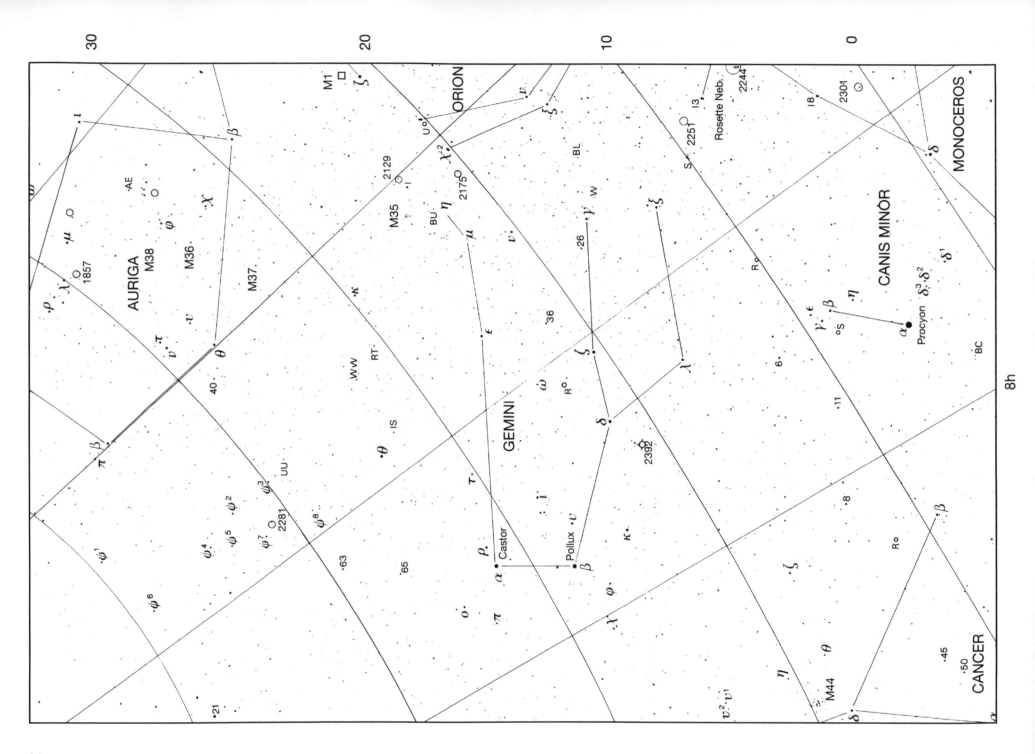

Map 18

MONOCEROS

This map contains the whole of Monoceros and Canis Minor, with parts of Orion, Canis Major, Puppis, Lepus, Gemini, Hydra and Pyxis. Of course Sirius is dominant; the orange hue of Theta Canis Majoris (4.1, K4) comes out well in the photograph, and so does the open cluster M41, near the bottom of the photograph below Sirius which looks decidedly blue even though visually it looks white. Other clusters in Canis Major can also be seen (NGC 2360, C58; NGC 2383 and NGC 2362) and are described with Map 19. The redness of the Orion Nebula, M42, is striking in the photograph; this again cannot be seen visually and is due to the fact that the main constituent of the nebular gas is hydrogen.

Monoceros adjoins Orion, and most of it is shown here, though it is also shown in Map 15. There are no bright stars in the celestial Unicorn, but there are various objects of interest. T Monocerotis is a Cepheid, not far from maximum (magnitude 6) at the time of the photograph; V Monocerotis is a Mira star, forming a triangle with Beta and Gamma, and was also near maximum, so that the redness is evident. X Monocerotis, a red semi-regular variable near the open cluster M50 at the bottom left of the map, was some way from its brightest, and is not well shown. M50 itself can easily be mistaken for a slightly brighter region of the Milky Way, but is definitely a bona-fide cluster with a diameter of about 13 light-years. It lies near Alpha Monocerotis (3.9, K0).

Beta Monocerotis is a triple star; the three components are not very unequal (magnitudes 4.7, 5.2 and 6.1), a rather unusual combination; it makes a fine sight in a telescope, and William Herschel, who discovered it in 1781, called it "one of the most beautiful sights in the heavens". In this region is the rather inconspicuous open cluster NGC 2215, not easy to trace in the photograph; its integrated magnitude is below 8, but it can just be made out as a tiny patch.

Perhaps the most spectacular object in Monoceros is the Rosette Nebula, NGC 2237–9 (C49), which surrounds the open cluster NGC 2244. The photograph shows it clearly, and it is said to be visible as a faint glow with binoculars; when imaged with adequate equipment it is truly glorious. Also in the area are the open cluster NGC 2251, which contains about 30 stars, and NGC 2264, the Cone Nebula round the irregular variable S Monocerotis; the Cone is not an easy object, and photography is needed to bring out its distinctive shape.

Hydra, the Watersnake, adjoins Monoceros, and part of the "head" is shown at the left of the photograph; Delta (4.2, A0), Sigma (4.4, K2) and Eta (4.3, B3), though the two brightest stars in the Head, Zeta (3.1, K0) and Epsilon (3.4, G0) are just off the photograph; they are shown in Map 29. The open cluster M48, which lies near the star Zeta Monocerotis (4.3, G2) is actually in Hydra; it is within binocular range, and can be seen on the photograph, though it is not particularly conspicuous. The position given for it by Messier was 4 degrees in error by declination, and it is now generally agreed that the object we call M48 is identical with NGC 2548, discovered by Messier himself in 1771. The little trio of stars of which C is the brightest member was formerly included in Monoceros, and the star now called F Hydræ (4.7, G4) used to be 31 Monocerotis.

Part of Puppis is shown here, including Rho (2.8, F6) and several open clusters, of which the most prominent are M93, M46 and M47. The whole of Puppis is shown in Map 20, and is described there.

MONOCEROS

Brightest stars

Star		Proper name	Mag.	Spectrum	RA h m s			Dec. ° ' "		
β	Beta		3.7 (combined)	B2+B3	06 28 48.9			−07 01 58		
	30		3.90	A0	08 25 39.5			−03 54 23		
α	Alpha		3.93	K0	07 41 14.8			−09 33 04		
γ	Gamma		3.98	K3	06 14 51.3			−06 16 29		

Variable stars

	RA h m	Dec. ° '	Range (mags.)	Type	Period (d)	Spectrum
V	06 22.7	−02 12	6.0–13.7	Mira	333.8	M
T	06 25.2	+07 05	6.0–6.6	Cepheid	27.02	F–K
S	06 41.0	+09 54	4.0–5.0?	Irreg.		07
X	06 57.2	−09 04	6.9–10.0	Semi-reg.	156	M
RY	07 06.9	−07 33	7.7–9.2	Semi-reg.	466	N
U	07 30.8	−09 47	6.1–8.1	RV Tauri	92.3	F–K

Double stars

	RA h m	Dec. ° '	PA °	Sep. sec	Mags.
ε	06 23.8	+04 36	027	13.4	4.5, 6.5
β	06 28.8	−07 02	AB 132	7.3	4.7, 5.2
			AC 124	10.0	6.1
			AD 056	25.9	12.2
S (15)	06 41.0	+09 54	AB 213	2.8	4.7v, 7.5
			AC 013	16.6	9.8
			AD 308	41.3	9.6
			AE 139	73.9	9.9
			AF 222	156.0	7.7
			AK 056	105.6	8.1

Open clusters

M/C	NGC	RA h m	Dec. ° '	Diameter min.	Mag.	No of stars
C50	2244	06 32.4	+04 52	24	4.8	100
(In Rosette Nebula)						
	2251	06 34.7	+08 22	10	7.3	30
	2286	06 47.6	−03 10	15	7.5	50
	2301	06 51.8	+00 28	12	6.0	80
M50	2323	07 03.2	−08 20	16	5.9	80
	2335	07 06.6	−10 05	12	7.2	35
	2343	07 08.3	−10 39	7	6.7	20
	2353	07 14.6	−10 18	20	7.1	30
C54	2506	08 00.2	−10 47	7	7.6	150

Nebulæ

M/C	NGC	RA h m	Dec. ° '	Dimensions min.	Mag. of illum. star
C49	2237–9	06 32.3	+05 03	80 × 60	
(Rosette Nebula)					
C46	2261	06 39.2	+08 44	2 × 1	10v R Monocerotis
	2264	06 40.9	+09 54	60 × 30	4v S Monocerotis
(Cone Nebula)					

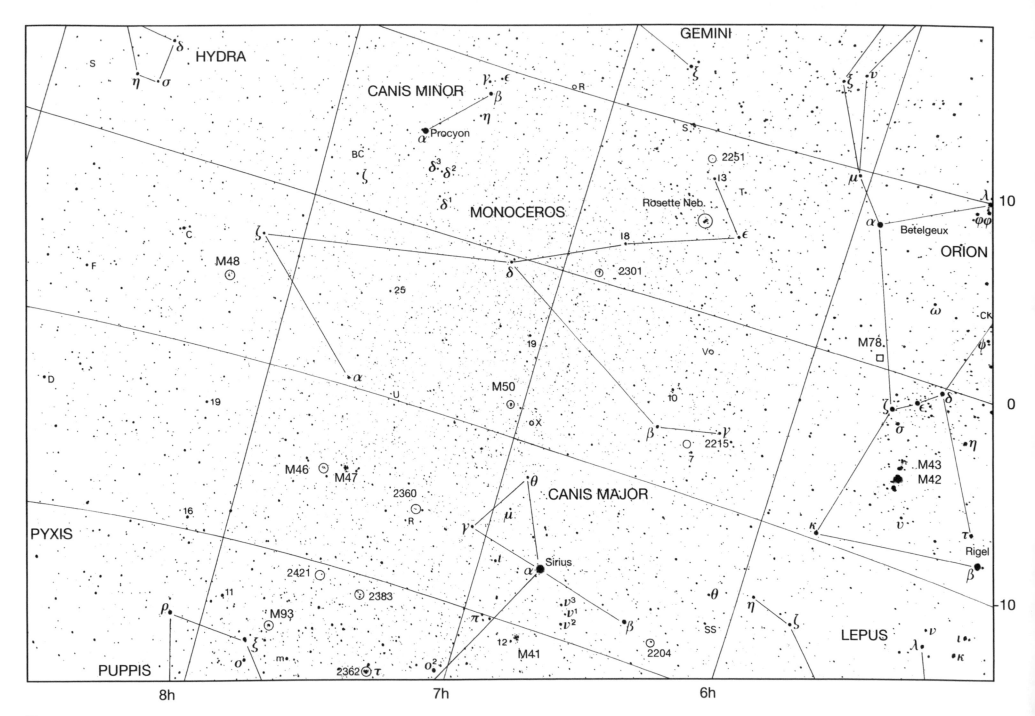

HYDRA

δ
S
η σ

CANIS MINOR
γ ε
β
η
α Procyon

GEMINI
ξ
ξ υ
μ

BC
ζ
δ³ δ²

δ¹

MONOCEROS

○ R
S
2251 ○
13
Rosette Neb.
T
ε

λ 10
α
Betelgeux
φφ
ORION

C
F
M48 ○

ζ
25

18
2301 ○
19
Vo
10

CK
ω

M78 □
ψ

D
19

α
U

M50 ○
○ X

β γ
○ 2215
7

0
δ
ζ ε
σ
η

M46 ○ M47

2360 ○
R

θ CANIS MAJOR
μ
γ
ι
α Sirius

M43
M42
κ
υ
τ
Rigel
β

PYXIS

16

2421 ○
11
2383 ○
M93 ○

ρ
ζ
o
m
2362 ○ τ
o²

π
12
M41

v³
v¹
v²
β

θ
SS
η ζ
LEPUS
λ
v
ι κ

PUPPIS

8h   7h   6h

2204 ○

90

Map 19

CANIS MAJOR

This map shows the whole of Canis Major and also Columba (described with Map 16), with parts of Monoceros, Lepus and Puppis.

Of course the dominant feature is Sirius, so much the most brilliant star in the sky – though it owes its eminence to the fact that cosmically speaking it is so near at hand; it is a mere 8 light-years away, and no more than 26 times as luminous as the Sun. It is interesting to compare it with Rigel. Sirius is more than a magnitude and a half the brighter of the two – but less than 1/2,000 as powerful!

Sirius is a pure white star; when low down it seems to flash various colours, but this is due entirely to the effects of the Earth's atmosphere through which the starlight has to pass. It is rather strange that some ancient observers, including Ptolemy, described it as red. It is certainly not red now, and it is not the sort of star which would be expected to show permanent change over a period of a few centuries. Any real alteration is unlikely to the highest degree; all the same, it is an intriguing little puzzle.

Sirius is about 2½ times as massive as the Sun, and has a diameter of around 2.7 million kilometres (1.7 million miles). It is not a solitary traveller in space; it has a white dwarf companion only 1/10,000 as bright – and since Sirius is often known as the Dog-Star, the companion has inevitably been nicknamed the Pup! It is the best-known example of a white dwarf; that is to say a bankrupt star, which has exhausted all its nuclear reserves, and is now very small and amazingly dense. In fact the diameter of the Pup is no more than 42,000 kilometres (26,000 miles) or so, but the mass is equal to that of the Sun, so that it is over 70,000 times as dense as water. It is

not particularly faint in our sky; its magnitude is 8.6, and if it could be seen shining on its own it would be an easy binocular object, but it is so overpowered by the glare of Sirius that it is very elusive. The separation is never more than 11.5 seconds of arc, and when the separation is least, as in 1995, it is reduced to only 3 seconds of arc. The revolution period is 50 years.

Some of the other stars of Canis Major are highly luminous; for example Delta could match 130,000 Suns, and is over 3,000 light-years away, while Epsilon or Adhara – only just below the official first magnitude – is exceptionally powerful at extreme ultra-violet wavelengths. Epsilon has a 7.4-magnitude companion at a separation of over 7 seconds of arc; the two seem to be genuinely associated, and share a common motion through space even though they are a very long way apart. Mu (5.3, M0) is a neat little double; the companion is white, giving a good colour contrast, and there are two fainter and more distant companions. The red tint of the primary shows up well in the photograph; the star lies inside the triangle of Sirius, Theta (4.1, K4) and Gamma (4.1, B8). (Though given the third letter of the Greek alphabet, Gamma is relatively faint; here, as in other constellations, the original Bayer method has not been followed!)

Several variables in Canis Major are shown in the photograph. R, to the left of Gamma on the map, is an Algol-type eclipsing binary with a magnitude range of 5.7 to 6.3; it is of type F, and is not noticeably coloured, so that there is nothing particular to single it out. W, near the top left of the map, is a red star of type N; it is a semi-regular, fluctuating between magnitudes 6.4 and 7.9

– at the time of the photograph the magnitude was about 7. UW, not far from Delta, is a massive eclipsing binary of what is termed the Beta Lyræ type; the components are much less unequal than with Algol stars, so that there are two pronounced minima occurring alternately, one deeper than the other. With UW the components are only about 31 million kilometres (19 million miles) apart, so that they cannot be seen separately, and gravitational forces must have distorted them into the shapes of eggs.

UW lies in the field of the Tau Canis Majoris cluster, NGC 2362 (C64). Tau itself is an O-type star 20,000 times as luminous as the Sun; the cluster is easily identifiable, and appears somewhat bluish. However, the most prominent open cluster in the constellation is M41, south of Sirius and the little trio of stars lettered Nu. M41 is easily visible with the naked eye; it is roughly circular, and in it there is one comparatively prominent red giant star. Its distance is around 1,600 light-years. The other open clusters shown in the photograph, NGC 2383 and 2204, are much less conspicuous. NGC 2204, in the region of Beta, contains at least 80 stars, but all are rather faint.

CANIS MAJOR

Brightest stars

Star	Proper name	Mag.	Spectrum	RA h m s	Dec. ° ' "
α Alpha	Sirius	−1.46	A1	06 45 08.9	−16 42 58
ε Epsilon	Adhara	1.50	B2	06 58 37.5	−28 58 20
δ Delta	Wezea	1.86	F8	07 08 23.4	−26 23 36
β Beta	Mirzam	1.98v	B1	06 22 41.9	−17 57 22
η Eta	Aludra	2.44	B5	07 24 05.6	−29 18 11
ζ Zeta	Phurad	3.02	B3	06 20 18.7	−30 03 48
o² Omicron²		3.03	B3	07 03 01.4	−23 50 00
o¹ Omicron¹		3.86	K3	06 54 07.8	−24 11 02
ω Omega		3.86	B3	07 14 48.6	−26 46 22
ν² Nu²		3.95	K1	06 36 41.0	−19 15 22
κ Kappa		3.96	B2	06 49 50.4	−32 30 31

Variable stars

	RA h m	Dec. ° '	Range (mags.)	Type	Period (d)	Spectrum
W	07 08.1	−11 55	6.4–7.9	Irreg.		N
UW	07 18.4	−24 34	4.0–5.3	Beta Lyræ	4.39	07
R	07 19.5	−16 24	5.7–6.3	Algol	1.14	F

Double stars

	RA h m	Dec. ° '	PA °	Sep. sec	Mags.	
α	06 45.1	−16 43	005	4.5	−1.5, 8.3	Binary, 50 y
μ	06 56.1	−14 03	340	3.0	5.3, 8.6	Isis
ε	06 58.6	−28 58	161	7.5	1.5, 7.4	

Open clusters

M/C	NGC	RA h m	Dec. ° '	Diameter min.	Mag.	No of stars
	2204	06 15.7	−18 39	13	8.6	80
M41	2287	06 47.0	−20 44	38	4.5	80
	2345	07 08.3	−13 10	12	7.7	20
C58	2360	07 17.8	−15 37	13	7.2	80
C64	2362	07 17.8	−24 57	8	4.1	60 τ CMa

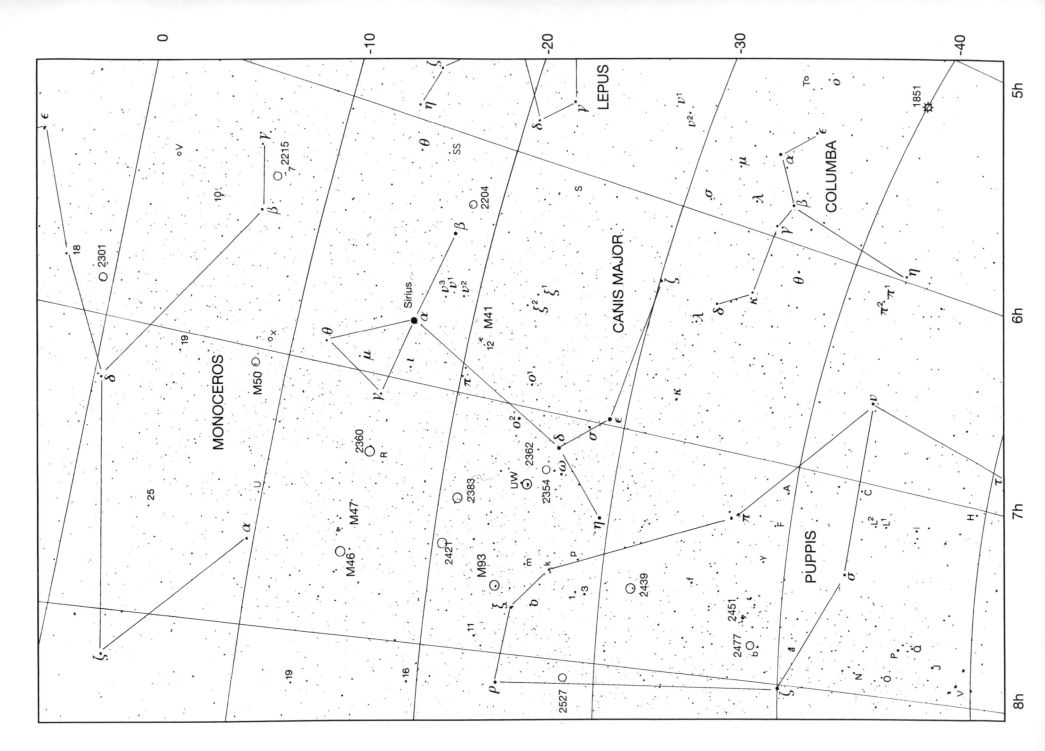

Map 20

PUPPIS, ANTLIA

Here, Argo is dominant; virtually the whole of Vela and Puppis, with Pyxis (originally Malus, the Mast) and a small section of Carina are shown. The whole area is very rich, and is crossed by the Milky Way. Carina is described with Map 22.

A very notable feature of this region is the False Cross; its four stars are Epsilon Carinæ (1.9, K0); Iota Carinæ (2.2, F0); Kappa Velorum (2.5, B2) and Delta Velorum (2.2, A0). Many people confuse it with the Southern Cross, and it is true that the shape is much the same; moreover, as with the Southern Cross, three of the stars are white or bluish-white and the fourth (in this case, Epsilon Carinæ) orange-red. However, the False Cross is much the larger and less brilliant of the two, and it is also more symmetrical. The difference between its brightest and faintest members is only about half a magnitude, while in the Southern Cross one of the four (Delta Crucis) is far inferior to the others. The photograph shows the False Cross well; compare it with the Southern Cross in Map 23.

Parts of Puppis have been shown on other maps, but most of the constellation is included here. The brightest star is Zeta (2.2, O5.8), a very powerful star about 60,000 times as luminous as the Sun, and therefore fully the equal of Rigel; it is about 2,400 light-years away. It is one of the very few bright stars to have an O-type spectrum.

(Incidentally, there are no stars in Puppis carrying the first five letters of the Greek alphabet. When the old, huge Argo Navis was cut up, Alpha, Beta and Epsilon went to Carina (the Keel) while Gamma and Delta were allotted to Vela (the Sails). Alpha Argus, now Alpha Carinæ, is of course Canopus.)

In 1942 a bright nova, CP Puppis, flared up in a position within five degrees of Zeta. It rose to magnitude 0.4, and is actually the brightest nova of the twentieth century apart from those in Perseus (1901) and Aquila (1918). However, CP Puppis did not remain bright for long; within a month it had dropped below naked-eye visibility, and it has now reverted to its pre-outburst magnitude of 17, so that it is far too faint to show up on this photograph. At its peak it must have sent out as much energy as Canopus, and it was at least 5,000 light-years away. Most galactic novæ appear in or near the Milky Way, and in this picture the Milky Way runs from the top right to the lower left, passing by Zeta.

One of the most interesting objects in Puppis is the red semi-regular variable L^2, near the reddish Sigma (3.2, K5). It makes up a pair with its neighbour L^1 (5.0, A0), but the two are not genuinely associated. L^2 is one of the brightest variables in the sky, since at maximum it can rise to magnitude 2.6; it is of type M, and there is a period of 140 days which is subject to marked irregularities. When this picture was taken it was of about magnitude 5, equal to L^1, but its orange hue is always unmistakable. At minimum it drops to the limit of naked-eye visibility. It is 75 light-years away, with a mean luminosity about 1,400 times that of the Sun.

The eclipsing binary V Puppis lies about eight degrees south east of L^2, near the bright star Gamma Velorum. V Puppis is a system of the Beta Lyræ type, made up of two B-type giants which are almost in contact and which mutually eclipse each other; the period is 10 hours 54 minutes 47 seconds, and the magnitude range is from 4.7 to 5.2. Suitable comparison stars, all identifiable on the photograph, are P (4.2, B0), J (4.3, B2), Q (4.6, G8) and O (5.2, M0).

Pi Puppis (2.7, K5) is orange. Very close to it are Upsilon¹ (4.7, B3) and Upsilon² (5.1, B3), which are bluish; with binoculars the colour contrast with Pi is attractive. The two Upsilons are probably associated with each other, since they have the same motion in space, but lie well beyond Pi, and must be at least 500 light-years away.

The open clusters in the northern part of Puppis (M47, M93, NGC 2421 and NGC 2383), shown on Map 19, are also visible on this photograph. Further south in the constellation there are several other notable clusters. NGC 2477 (C71), near Zeta, is very rich, and is a fine object in even a small telescope; NGC 2451, close beside it has a brighter integrated magnitude, but is looser and less spectacular. NGC 2439, round R Puppis, is different, and its nature is rather uncertain. It lies almost in line with Delta and Eta Canis Majoris, and is not hard to identify, but it has sometimes been classed as an asterism (a chance grouping of stars) rather than a true cluster. R Puppis itself (6.6, G2) has a variable star designation, but its variability is in some doubt, and at best the range can be no more than a tenth of a magnitude.

Sigma Puppis has a companion of magnitude 9.4; the two components have a common motion in space, but are a long way apart. It is interesting to note that the fainter member of the pair is just about equal to our Sun in luminosity, while the red primary is at least 120 times more powerful.

Antlia, at the top left of the map, covers almost 240 square degrees, but is very barren of interesting objects, though it is worth noting that two of its leading stars are red: Alpha (4.2, M0) and Epsilon (4.5, also M0). Of the two, the colour of Alpha is the more pronounced. Alpha makes up a little pair with Delta (5.6, A2) which is an easy double. Theta (4.8, G7) is also double, but since the separation is only 0.1 of a second of arc it is a very difficult object.

Two variable stars in Antlia are identifiable here. S Antliæ, slightly below and to the right of Theta, is an eclipsing binary of the W Ursæ Majoris type; it has a small range (6.4 to 6.9) and a period of only 16 hours. It is one of a line of faint stars, and it is interesting to follow its very quick changes even though they are not pronounced. The other variable is U Antliæ, well below and to the left of the Alpha-Delta pair, which is rather isolated and is easy to identify because of its colour. Its range is from magnitude 5.7 to 6.8, but its type is uncertain; in some lists it is given as a semi-regular (though no semblance of a period seems to have been established), while others, probably with greater justification, list it as irregular. Its N-type spectrum means that it is exceptionally red, and this comes out well in the photograph.

As may be gathered from its name, Antlia is a modern constellation; it was added to the sky by La Caille in 1752 as Antlia Pneumatica, the Airpump.

PUPPIS

Brightest stars

Star		Proper name	Mag.	Spectrum	RA h m s	Dec. ° ' "
ζ	Zeta	Suhail Hadar	2.25	O5.8	08 03 35.0	−40 00 12
π	Pi		2.70	K5	07 17 08.5	−37 05 51
ρ	Rho	Turais	2.81v	F6	08 07 32.6	−24 18 15
τ	Tau		2.93	K0	06 49 56.1	−50 36 53
ν	Nu		3.17	B8	06 37 45.6	−43 11 45
σ	Sigma		3.25	K5	07 29 13.8	−43 18 05
ξ	Xi	Asmidiske	3.34	G3	07 49 17.6	−24 51 35
	L²		3.4 max.	M5	07 13 13.3	−45 10 59
	c		3.59	cK	07 45 15.2	−37 58 07
	a		3.73	G5	07 52 13.0	−40 34 33
	k		3.82	B8	07 38 49.7	−26 48 13
	3		3.96	A2	07 43 48.4	−28 57 18

Variable stars

	RA h m	Dec. ° '	Range (mags.)	Type	Period (d)	Spectrum
L²	07 13.5	−44 39	2.6–6.2	Semi-reg.	140	M
Z	07 32.6	−20 40	7.2–14.6	Mira	499.7	M
W	07 46.0	−41 12	7.3–13.6	Mira	120.1	M
V	07 58.2	−49 15	4.7–5.2	Beta Lyræ	1.45	B+B
RS	08 13.1	−34 35	6.5–7.6	Cepheid	41.39	F–G

Open clusters

M/C	NGC	RA h m	Dec. ° '	Diameter min.	Mag.	No of stars
M47	2422	07 36.6	−14 30	30	4.4	30
	Mel 71	07 37.5	−12 04	9	7.1	80
	2439	07 40.8	−31 39	10	6.9	80
						R Puppis. Asterism
M46	2437	07 41.8	−14 49	27	6.1	100
M93	2447	07 44.6	−23 52	22	6.2	80
	2451	07 45.4	−37 58	45	2.8	40
C71	2477	07 52.3	−38 33	27	5.8	160
	2527	08 05.3	−28 10	22	6.5	40
	2533	08 07.0	−29 54	3.5	7.6	60
	2539	08 10.7	−12 50	22	6.5	50
	2546	08 12.4	−37 38	41	6.3	40
	2567	08 18.6	−30 38	10	7.4	40
	2571	08 18.9	−29 44	13	7.0	30

ANTLIA

No star above magnitude 4

Variable stars

	RA h m	Dec. ° '	Range (mags.)	Type	Period (d)	Spectrum
S	09 32.3	−28 38	6.4–6.9	W Uma	0.65	A
U	10 35.2	−39 34	5.7–6.8	Irreg.		N

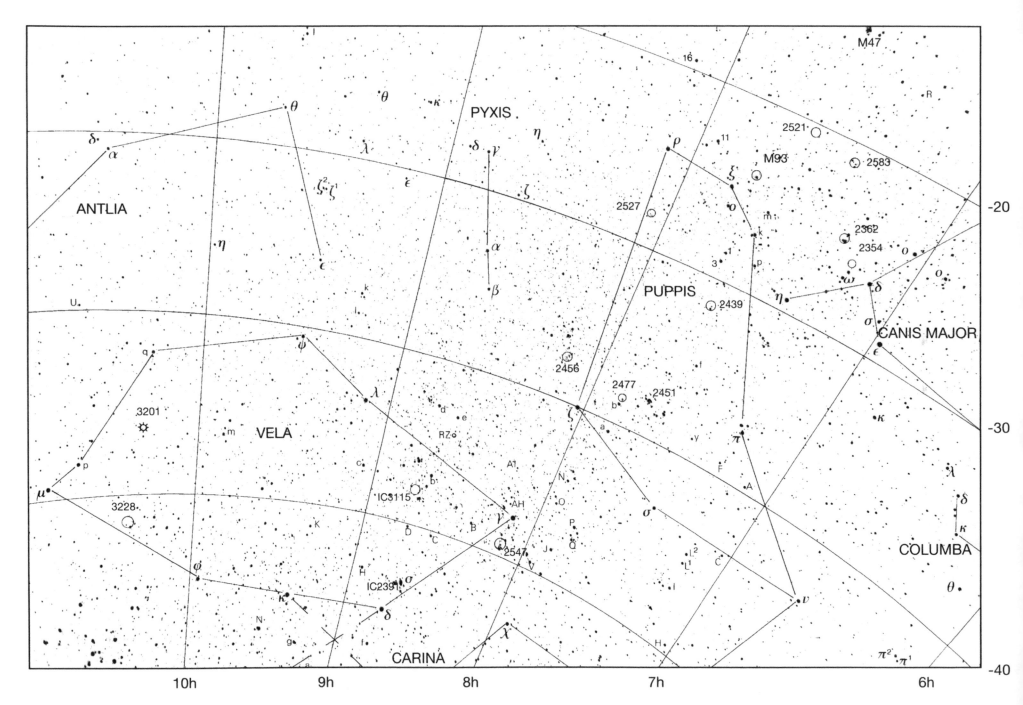

PYXIS

ANTLIA

PUPPIS

CANIS MAJOR

VELA

COLUMBA

CARINA

M47

M93

2521

2583

2527

2362

2354

2439

2456

2477

2451

3201

3228

2547

IC3115

IC239

-20

-30

-40

10h

9h

8h

7h

6h

Map 21

VELA, PYXIS, PICTOR

This map contains much of the old Argo Navis; almost the whole of Vela, a large part of Puppis and most of Carina, including Canopus. Pictor and Pyxis are included, and parts of Columba, Antlia, Dorado and Volans. Most of these stars are invisible from European latitudes; for example, Eta Columbæ in the southern part of the Dove does not rise from Britain and the northern United States and Canada, though the other main stars of the constellation do appear briefly over the southern horizon. Of course Canopus is dominant; note also the False Cross, which is partly in Carina and partly in Vela. Carina is described with Map 22.

The brightest star in Vela is Gamma, sometimes still known by its old proper name of Regor. It is of magnitude 1.8, and the spectrum is of the WC or Wolf-Rayet type; it is in fact the brightest member of this class. Wolf-Rayet stars, first described in 1867 by G. Wolf and G. Rayet at the Paris Observatory, show bright emission lines in their spectra, and are luminous and unstable, constantly ejecting envelopes of material. Gamma Velorum is 3,800 times as luminous as the Sun, and 520 light-years away. It has been suspected of being a very close binary, and certainly it is a splendid visual double; the B3-type companion is only just below the fourth magnitude, and since the separation is over 40 seconds of arc this is a particularly easy telescopic pair. Several other giant stars lie nearby. Delta Velorum, in the False Cross, is also an easy visual double; here the companion is of the fifth magnitude, and here too there are several close stars in the same telescopic field, which may well be related to the main pair.

Of the other leaders of Vela, Lambda (2.2, K5) is orange-red; the colour comes out well in the photograph. W Velorum, above Lambda in the map, is of spectral type F8, and has a variable star designation, but seems to be constant at magnitude 4.4. Its orange colour is fairly marked. Another orange star once suspected of being variable is N Velorum, which lies below Kappa in the False Cross. It is of type K5, and appears to shine steadily at magnitude 3.1.

Not far from Gamma Velorum there are three faint variables: AH and RZ (Cepheids) and AI (a Delta Scuti pulsating star). All are identifiable in the photograph. AI has a very short period (2 hours 41 minutes) and a range of from magnitude 6.4 to 7.1, so that its changes are obvious telescopically after a very brief observing session.

In the same low-power binocular field with Delta Velorum is the attractive open cluster IC 2391 (C85), round o Velorum. It is easily visible with the naked eye, and shows up well in the photograph; telescopically it has a somewhat cruciform appearance. IC 2395 is a small but fairly bright open cluster; NGC 3201, not far from q Velorum (3.8, A2) is just about traceable in the photograph, though it contains no more than about twenty stars. The faint little trio just below it on the map is a help in locating it. There is also a fairly prominent globular cluster, NGC 3228, roughly between Psi (3.6, F2 and Mu (2.7, G5). In the photograph it looks very like a faint star.

The Gum Nebula – the remnant of a supernova which blazed up in prehistoric times, and must have been truly spectacular – is in Vela, at RA 8h 30m declination –45°. It is of tremendous interest, and contains a pulsar, but unfortunately its low surface brightness means that it is not visible in the photograph.

Pyxis is an obscure constellation, but it does contain some objects worth noting. Zeta (4.9, G4) is a wide, easy telescopic double, and Kappa (4.6, M0) has a companion above the tenth magnitude; the primary is distinctly red. TY Pyxidis, near the little pair of Gamma (4.0, K4) and Delta (4.9, A3) is an eclipsing binary with a range of over half a magnitude (6.9 to 7.5). However, the most interesting object is the recurrent nova T Pyxidis. Generally it remains below magnitude 13, but on several occasions – as in 1890, 1902, 1944 and 1966 – it has brightened up slowly and come within binocular range; in 1966 it peaked at magnitude 6.3, and then slowly faded back to normal at the rate of about a magnitude a month. It lies not far from Epsilon (5.5, A3). Since the outbursts are quite unpredictable, the star is worth monitoring. Its distance is uncertain; there have been suggestions that it may be a very close binary.

Pictor, the Painter, adjoins Carina in the region of Canopus. Much the most interesting object is Beta, of magnitude 3.8; according to the Cambridge catalogue it is 78 light-years away and 58 times as luminous as the Sun. Outwardly there is nothing remarkable about it, but in 1983 the infra-red astronomical satellite IRAS discovered that it is associated with a mass of cool material, and this has since been confirmed visually; it may well be that this material is planet-forming, and it is not impossible that Beta Pictoris may be the centre of a true planetary system. Its spectral type is A5, so that as well as being more powerful than the Sun it is also hotter.

VELA

Brightest stars

Star		Proper name	Mag.	Spectrum	RA h m s	Dec. ° ' "
γ	Gamma	Regor	1.78	WC7	08 09 31.9	−47 20 12
δ	Delta	Koo She	1.96	A0	08 44 44.2	−54 42 30
λ	Lambda	Al Suhail al Wazn	2.21v	K5	09 07 59.7	−43 25 57
κ	Kappa	Markeb	2.50	B2	09 22 06.8	−55 00 38
μ	Mu		2.69	G5	10 46 46.1	−49 25 12
	N		3.13	K5	09 31 13.3	−57 02 04
φ	Phi		3.54	B5	09 56 51.7	−54 34 03
ψ	Psi		3.60	F2	09 30 41.9	−40 28 00
o	Omicron		3.62v	B3	08 40 17.6	−52 55 19
	c		3.75	K2	09 04 09.2	−47 05 02
	p		3.84	F4	10 37 18.0	−48 13 32
	b		3.84	F2	08 40 37.6	−46 38 55
	q		3.85	A2	10 14 44.1	−42 07 19
	a		3.91	A0	08 46 01.7	−46 02 30

Variable stars

	RA h m	Dec. ° '	Range (mags.)	Type	Period (d)	Spectrum
AI	08 14.1	−44 34	6.4–7.1	Delta Scuti	0.11	A–F
RZ	08 37.0	−44 07	6.4–7.6	Cepheid	20.40	G
CV	09 00.6	−51 33	6.5–7.3	Algol	6.89	B+B

Double stars

	RA h m	Dec. ° '	PA °	Sep. sec	Mags.	
γ	08 09.5	−47 20	AB 220	41.2	1.9, 4.2	
			AC 151	62.3	8.2	
			AD 141	93.5	9.1	
			DE 146	1.8	12.5	
δ	08 44.7	−54 43	AB 153	2.6	2.1, 5.1	
			AC 061	69.2	11.0	
			CD 102	6.2	13.5	
μ	10 46.8	−49 25	055	2.3	2.7, 6.4	Binary, 116 y

Open clusters

M/C	NGC	RA h m	Dec. ° '	Diameter min.	Mag.	No of stars
	2547	08 10.7	−49 16	20	4.7	80
	IC 2391	08 40.2	−53 04	50	2.5	30
(o Velorum Cluster)						
	IC 2395	08 41.1	−48 12	8	4.6	40
	2669	08 44.9	−52 58	12	6.1	40
	2670	08 45.5	−48 47	9	7.8	30
	IC 2488	09 27.6	−56 59	15	7.4	70
	2910	09 30.4	−52 54	5	7.2	30
	2925	09 33.7	−53 26	12	8.3	40
	3228	10 21.8	−51 43	18	6.0	15

Globular cluster

M/C	NGC	RA h m	Dec. ° '	Diameter min.	Mag.
	3201	10 17.6	−46 25	18.2	6.7

PYXIS

Brightest stars

Star		Proper name	Mag.	Spectrum	RA h m s	Dec. ° ' "
α	Alpha		3.68	B2	08 43 35.5	−33 11 11
β	Beta		3.97	G4	08 40 06.1	−35 18 30

Variable stars

	RA h m	Dec. ° '	Range (mags.)	Type	Period (d)	Spectrum
TY	08 59.7	−27 49	6.9–7.5	Eclipsing	3.20	G+G
T	09 04.7	−32 23	6.3–14.0	Recurrent nova		Q

(Outbursts 1890, 1902, 1920, 1944, 1966)

Double star

	RA h m	Dec. ° '	PA °	Sep. sec	Mags.
ζ	08 39.7	−29 34	061	52.4	4.9, 9.1

PICTOR

Brightest stars

Star		Proper name	Mag.	Spectrum	RA h m s	Dec. ° ' "
α	Alpha		3.27	A5	06 48 11.4	−61 56 29
β	Beta		3.85	A5	05 47 17.1	−51 03 59

Variable stars

	RA h m	Dec. ° '	Range (mags.)	Type	Period (d)	Spectrum
R	04 46.2	−49 15	6.7–10.0	Semi-reg.	164	M
S	05 11.0	−48 30	6.5–14.0	Mira	426.6	M

Double star

	RA h m	Dec. ° '	PA °	Sep. sec	Mags.
τ	04 50.9	−53 28	058	12.3	5.6, 6.4

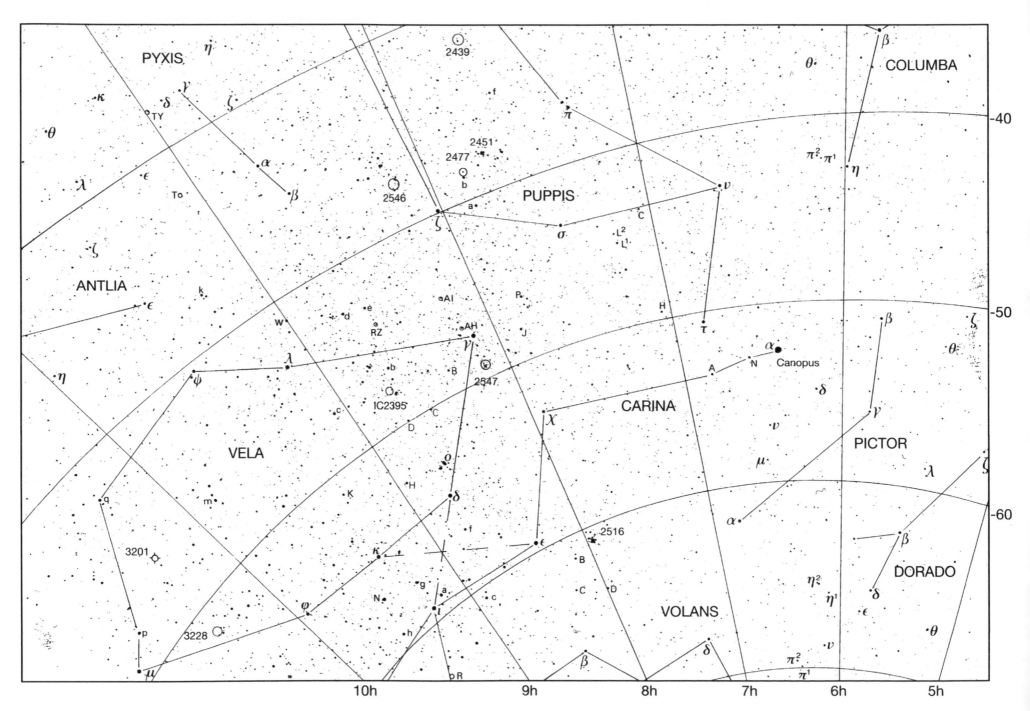

Map 22

CARINA, VOLANS

This map includes the whole of Carina, Pictor, Volans and Mensa, with most of Vela and parts of Dorado, Centaurus, Crux and Chamæleon.

The four brightest stars shown here give a good indication of the magnitude scale in the maps. They are Canopus (–0.7), upper right; Epsilon Carinæ (1.9) just above centre; Beta Carinæ (1.7) below centre, and Alpha Crucis (0.8), lower left corner. Visually, Beta Carinæ is slightly brighter than Epsilon, but in this photograph Epsilon looks the more prominent of the two. This is because of a marked difference in colour. Epsilon, in the False Cross, is an orange K-type star, while Beta is of type A, and is pure white. The False Cross is prominent; note the apparent blueness of the cluster NGC 2516 (C96) to the right of Epsilon.

Two objects stand out at once. One is of course Canopus, which far outshines every other star in the sky apart from Sirius, and according to the Cambridge catalogue is 200,000 times as luminous as the Sun (against a mere 26 Sun-power for Sirius). The other is the nebulosity associated with Eta Carinæ, the most erratic of all variables – and at times the most spectacular. For a while, during the last century, it outshone even Canopus, though for more than a hundred years now it has been below naked-eye visibility. In its heyday it was known as Eta Argûs; this was before the Ship had been dismantled!

Eta Carinæ is unique. Our first record of it seems to be due to Edmond Halley, who went to St Helena in 1677 to make the first systematic survey of the far-southern sky; he reported that it was of the fourth magnitude. From then until 1820 it fluctuated between magnitudes 2 and 4, but then began to brighten, and by 1827 had reached the first magnitude. Then came its period of glory. In 1837 Sir John Herschel, observing from the Cape, ranked it equal to Alpha Centauri. After a temporary fade, it burst forth, and in 1843 was almost as brilliant as Sirius. Slowly a decline set in, and by 1870 it had dropped below naked-eye range. Since then it has hovered between magnitudes 6 and 7.5.

It is likely that Eta Carinæ is the most powerful star known; at its peak it may have been 6 million times as luminous as the Sun, so that in comparison even Canopus appears puny. Eta is also very massive – perhaps over 100 times as massive as the Sun. This is very near the limit of instability, and Eta is indeed violently unstable; during its outburst in the 1840s it seems to have ejected massive amounts of gas and dust moving at speeds of almost 3 million kilometres (1.9 million miles) per hour. The associated nebulosity – the Keyhole Nebula NGC 3372 (C92) – has been beautifully imaged with the Hubble Space Telescope, and a tenuous outer cloud has been revealed, made up of the quickest-moving ejected gas. The inner portion of two roughly spherical expanding lobes, made of dust which is being illuminated by the star itself.

Eta Carinæ is almost as powerful today as it was in the 1840s; it appears dimmer partly because it is more strongly veiled by nebulosity, and partly because its main emission is in infra-red. Through a telescope it looks quite unlike a normal star, and is more like an "orange blob". The colour is very marked in the photograph. Of course other nebulæ are shown as red – M42, in Orion's Sword, being a classic example – but in most cases the redness is not visible with the naked eye or telescopically, whereas with Eta Carinæ it cannot be missed, and contrasts sharply with the apparent blueness of nearby open clusters such as NGC 3572, 3114 and IC 2602 (C102).

Eta Carinæ is so massive, and so unstable, that it must eventually explode as a supernova, and even from its distance of at least 6,000 light-years it will then be imposing by any standards. Just when this will happen we do not know – but happen one day, it surely must.

NGC 2516 (C96), close to Epsilon, is a magnificent open cluster; it is very prominent in the photograph, and is easily visible with the naked eye. Another naked-eye cluster is IC 2602 (C102), round Theta Carinæ (2.8, B0); its distance is of the order of 750 light-years.

A reddish star in the same region is q Carinæ (3.4, K5); the Mira variable S Carinæ is visible slightly above and to the right of q, and this too is red, though when the photograph was taken it was some way from its maximum brightness and was of about magnitude 7 (at its best it can attain 4.5). Another Mira star is R Carinæ. it is seen here almost directly below Iota in the False Cross, slightly above and to the right of 1 Carinæ (4.0, F2). It can be among the brightest of the Mira stars, rising to magnitude 3.9, but when this picture was taken it was not much above the eighth magnitude, so that its redness is barely detectable. The little quadrilateral of stars close beside it, to the upper right, is a help in identification.

The globular cluster NGC 2808 can just be made out, and is easier to see on this photograph than on the image with Map 10. It is in Carina though the nearest fairly bright star is Alpha Volantis (4.0, A5). The integrated magnitude is only just below 6, so that it is an easy telescopic object.

Volans intrudes into Carina, between Canopus and Beta Carinæ (Beta appears at the lower left of the map). In Volans we have several doubles: Gamma, Zeta, Epsilon and Kappa. Gamma is only just too close to be split with binoculars. Finally, there is Chamæleon, with no star above the fourth magnitude; the brightest is Alpha (4.1, F6). Alpha makes a neat little pair with Theta (4.4, K0).

Beta Pictoris, probably the best-known candidate as the centre of a system of planets, shows up to the right of Canopus. The Large Cloud of Magellan is also visible, between Delta Doradûs (4.3, A7) and Beta Mensæ (5.3, G8), but it does not show up well, though with the naked eye it is extremely prominent – even moonlight will not drown it – and it is much the brightest of the external galaxies. It is much smaller than our Galaxy, at a distance of 169,000 light-years, and ranks as a satellite system. It was once classed as irregular, but there are obvious indications of a barred spiral structure. It contains objects of all kinds, including the magnificent Tarantula Nebula, NGC 2070 (C103); if the Tarantula were as close to us as the Orion Nebula, it would cast shadows. In 1987 a supernova blazed out in the Large Cloud, temporarily transforming the aspect of the whole of that part of the sky.

A small part of Centaurus is shown at the lower left, including the Lambda Centauri cluster. Alpha Crucis is just in the frame. Note the neat little pair made up of Theta[1] Crucis (4.5, A5) and Theta[2] (5.0, B3).

CARINA

Brightest stars

Star	Proper name	Mag.	Spectrum	RA h m s	Dec. ° ' "
α Alpha	Canopus	−0.72	F0	06 23 57.1	−52 41 44
β Beta	Miaplacidus	1.68	A0	09 13 12.2	−69 43 02
ε Epsilon	Avior	1.86	K0	08 22 30.8	−59 30 34
ι Iota	Tureis	2.25	F0	09 17 05.4	−59 16 31
θ Theta		2.76	B0	10 42 57.4	−64 23 39
υ Upsilon		2.97	A7	09 47 06.1	−65 04 18
ρ Rho		3.32v	B3	10 32 01.4	−61 41 07
ω Omega		3.32	B7	10 13 44.3	−70 02 16
q		3.40	K5	10 17 04.9	−61 19 56
a		3.44	B2	09 10 57.9	−58 58 01
χ Chi		3.47	B2	07 56 46.7	−52 58 56
u		3.78	K0	10 53 29.6	−58 51 12
R		3.8 max.	M5	09 32 14.7	−62 47 19
c		3.84	B8	08 55 02.8	−60 38 41
x		3.91	G0	11 08 35.3	−58 58 30

Variable stars

	RA h m	Dec. ° '	Range (mags.)	Type	Period (d)	Spectrum
R	09 32.2	−62 47	3.9–10.5	Mira	308.7	M
ZZ	09 45.2	−62 30	3.3–4.2	Cepheid	35.53	F–K (l Carinæ)
S	10 09.4	−61 33	4.5–9.9	Mira	149.5	K–M
η	10 45.1	−59 41	−0.8–7.9	Irreg.		Pec.
U	10 57.8	−59 44	5.7–7.0	Cepheid	38.77	F–G

Double star

	RA h m	Dec. ° '	PA °	Sep. sec	Mags.
u	09 47.1	−65 04	127	5.0	3.1, 6.1

Open clusters

M/C	NGC	RA h m	Dec. ° '	Diameter min.	Mag.	No of stars
C96	2516	07 58.3	−60 52	30	3.8	80
	3114	10 02.7	−60 07	35	4.2	
	IC 2581	10 27.4	−57 38	8	4.3	25
C102	IC 2602	10 43.2	−60 24	50	1.9	60
(θ Carinæ Cluster)						
C91	3532	11 06.4	−58 40	55	3.0	150
	IC 2714	11 17.9	−62 42	12	8.2	100
	3680	11 25.7	−43 15	12	7.6	30

Globular cluster

M/C	NGC	RA h m	Dec. ° '	Diameter min.	Mag.
	2808	09 12.0	−64 52	13.8	6.3

Nebula

M/C	NGC	RA h m	Dec. ° '	Dimensions min.	Mag. of illum. star
C92	3372	10 43.8	−59 52	6.2	var.
(η Carinæ Nebula)					

VOLANS

Brightest stars

Star	Proper name	Mag.	Spectrum	RA h m s	Dec. ° ' "
γ Gamma		3.6 (combined)	G8+dF4	07 08 43	−70 29 54
β Beta		3.77	K2	08 25 44.3	−66 08 13
ζ Zeta		3.95	K0	07 41 49.3	−72 36 22
δ Delta		3.98	F8	07 16 49.8	−67 57 27

Double stars

	RA h m	Dec. ° '	PA °	Sep. sec	Mags.
γ²	07 08.8	−70 30	300	13.6	4.0, 5.9
ζ	07 41.8	−72 36	116	16.7	4.0, 9.8
ε	08 07.9	−68 37	024	6.1	4.4, 8.0
κ	08 19.8	−71 31	AB 057	65.0	5.4, 5.7
			BC 030	37.7	8.5

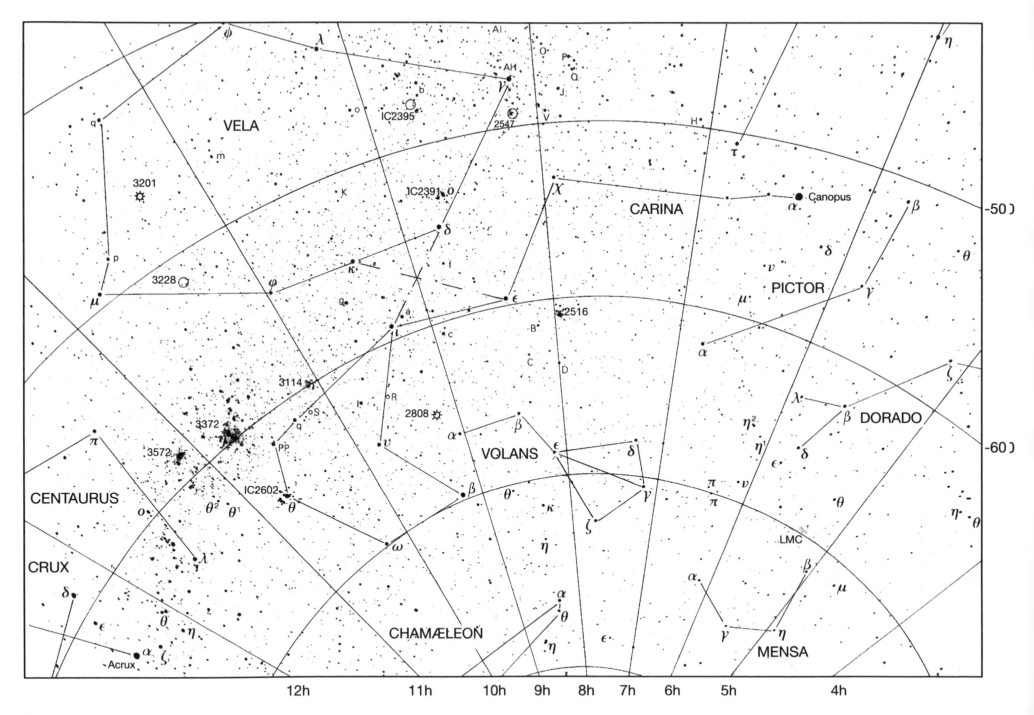

VELA

3201

3228

q

p

μ

φ

m

K

IC2395

λ

ψ

AI

AH

γ

O

P

Q

J

V

2547

o

IC2391 o

δ

κ

f

χ

ε

H

CARINA

τ

ι

α Canopus

β

-50

δ

θ

v

PICTOR

μ

γ

α

2516

B

C

D

ζ

λ

β DORADO

δ

η²

η¹

ε

δ

v

PP

3114

3372

η

3572

π

o

IC2602

θ² θ¹

θ

a

ι

c

R

S

q

v

2808

α

β

ω

β

θ

κ

ε

VOLANS

ε

δ

γ

ζ

η

π

π

v

LMC

θ

η θ

α

β

μ

CENTAURUS

CRUX

δ

ε

θ

η

α ζ

Acrux

λ

α

θ

CHAMÆLEON

η

ε

η

γ

η

MENSA

-60

12h 11h 10h 9h 8h 7h 6h 5h 4h

Map 23
CRUX AUSTRALIS

Unquestionably this is one of the finest regions in the sky; it contains the Eta Carinæ complex, the Southern Cross, and some of the richest regions of the Milky Way. Northern-hemisphere observers never cease to regret that it is inaccessible from the latitudes of northern Europe, the northern United States and Canada, but it is a mistake to suppose, as many people do, that to see the Southern Cross one has to travel over the equator. The declination of Alpha Crucis, the brightest star of the Cross, is –63°, so that in theory it can rise from anywhere south of latitude +27° (90 minus 63 = 27). Thus Alpha Crucis rises from Hilo in Hawaii (latitude +20°) but not from Cairo in Egypt (latitude +30°); it is circumpolar from Sydney and Cape Town (latitude –34°) but not from Johannesburg (–26°), where it dips briefly below the horizon. Interestingly, Crux was not even recognized as a separate constellation until 1679, when it was granted independence by an otherwise obscure astronomer named Augustin Royer; previously it had been included in Centaurus, which almost surrounds it.

The Eta Carinæ nebula is shown slightly to the left of the centre of the photograph, and the star-clouds and dark dusty regions stand out with remarkable clarity. The blue cast of NGC 3114, to the right of Eta Carinæ, is very marked even though it is not obvious when the cluster is viewed telescopically. The Theta Carinæ cluster, IC 2602 (C102) also looks bluish, and is very prominent indeed; the arrangement of the leading stars slightly recalls that of the Pleiades, and indeed the cluster has been nicknamed "the Southern Pleiades". The nebulosity round the Lambda Centauri cluster (IC 2944, C100), below and to the left of Eta Carinæ, shows up as red.

The faint stars of the Milky Way are better shown than in the Map 22 picture, and the areas of nebulosity seem more extended. The open cluster NGC 2516 (C96) in Carina is very striking to the mid-right of the photograph, below and to the right of Epsilon Carinæ, the orange-red star of the False Cross. Again it is interesting to compare the two Crosses; note the brilliance of Crux against its larger, fainter imitator. Beta Centauri, the second of the Pointers to Crux, appears at the extreme edge of the photograph (bottom left); midway up the left-hand edge is Gamma Centauri (2.2, A0) with its orange neighbour e Centauri (4.3, K5).

The Mira variable R Carinæ is shown here, between Iota (in the False Cross) and Upsilon (3.0, A7), but as in the Map 22 picture it was not far from minimum at the time. It is easy to trace because of its nearness to a little line of three stars, the lower of which is double. At its best, R Carinæ may rise above the fourth magnitude.

Crux is the smallest of all the constellations, covering only 88 square degrees, but it is exceptionally rich. In shape it is not cruciform, but is much more like a kite; it is a pity that there is no central star to make up a proper X! Moreover, it is not symmetrical. The four leaders are Alpha (0.8, B1+B3); Beta (1.2, B3); Gamma (1.6, M3) and Delta (2.8, B2), so that the main pattern is spoiled by the fact that Delta is more than a magnitude fainter than any of the other three. The symmetry is further disturbed by the presence of Epsilon (3.6, K2), which is orange. Of course the stars in the pattern are not genuinely associated with each other, and it is interesting to give their distances in light-years; the figures are 59 for Epsilon, 88 for Gamma, 260 for Delta, 360 for Alpha

and 425 for Beta. Beta, shining more than 8,000 times as powerfully as the Sun, is much the most luminous. It is a variable of the Beta Canis Majoris type, pulsating regularly, but the magnitude range is so small that the fluctuations are quite undetectable except with sensitive equipment.

Alpha Crucis (Acrux) is one of the finest doubles in the sky. Its dual nature was first noticed in 1685 by a French Jesuit priest, Father Guy Tachard, who was on his way to Siam (presumably to convert the natives to Christianity) and called in at Cape Town en route, where he set up a temporary observatory. The components of Alpha Crucis are not so very unequal, at magnitudes 1.4 and 1.9, and since the separation is well over 4 seconds of arc almost any telescope will split the pair. Both are of type B, and respectively 3,700 and 2,000 times as luminous as the Sun. Undoubtedly they are physically associated, but the real separation between them is of the order of 500 astronomical units – so that the orbital revolution period must be very long indeed.

There are three Cepheids in Crux – R, S and T – all of which have small ranges of between magnitudes 6 and 7, and all of which can be identified in the photograph. S, not far from Mu, is the easiest to find. Both R and T were near minimum when the picture was taken; they lie closely above Alpha, and between them is a rather inconspicuous open cluster, NGC 4349, which in the photograph looks very like a star.

The glorious cluster round Kappa Crucis, NGC 4755 (C94) is often called the Jewel Box. It lies close to Beta, and is easy to see with the naked eye as a hazy patch; binoculars show it well, and with a telescope it is superb.

Three of its leading stars form a triangle, and are hot and white or bluish; within the triangle is a red supergiant – probably as luminous as Betelgeux – which stands out at once, and led Sir John Herschel, in the 1830s, to give the cluster its familiar nickname. In the photograph the cluster appears more or less stellar, but with even a small telescope it ranks as one of the most magnificent sights in the sky. Other open clusters, such as NGC 4609 (C98), close to Alpha, and NGC 4103, below and to the right of Delta, seem pale by comparison.

Yet perhaps the most remarkable of all the objects in Crux is the Coal Sack (C99), the best-known of all dark nebulæ. It measures 7 by 5 degrees, and shows up as a virtually starless patch adjoining Alpha and Beta. It was noticed by Portuguese sailors during the sixteenth century, though no doubt it had been seen at a much earlier period. It is a dark mass, probably 500 to 600 light-years away, with a real diameter of around 7 light-years. It is very efficient at blocking out the light from objects beyond (one can understand why Sir William Herschel once referred to dark nebulæ as "holes in the heavens"!), but it is important to remember that a dark nebula differs from a bright one only because it is not being lit up by adjacent stars. It may well be that the other side of the Coal Sack is illuminated, so that if we could observe it from a different vantage point in the Galaxy it would appear bright. Many other dark nebulæ and dark rifts in the Milky Way are known, but the Coal Sack is much the largest and best defined of them.

CRUX AUSTRALIS

Brightest stars

Star	Proper name	Mag.	Spectrum	RA h m s	Dec. ° ' "
α Alpha	Acrux	0.83	B1+B3	12 26 35.9	−63 05 56
β Beta		1.25v	B0	12 47 43.2	−59 41 19
γ Gamma		1.63	M3	12 31 09.9	−57 06 47
δ Delta		2.80	B2	12 15 08.6	−58 44 55
ε Epsilon		3.59	K2	12 21 21.5	−60 24 04

Variable stars

	RA h m	Dec. ° '	Range (mags.)	Type	Period (d)	Spectrum
T	12 21.4	−62 17	6.3–6.8	Cepheid	6.73	F
R	12 23.6	−61 38	6.4–7.2	Cepheid	5.83	F–G
S	12 54.4	−58 26	6.2–6.9	Cepheid	4.69	F–G

Double stars

	RA h m	Dec. ° '	PA °	Sep. sec	Mags.
α	12 26.6	−63 06	AB 115	4.4	1.4, 1.9
			AC 202	90.1	1.0, 4.9
γ	12 31.2	−57 07	AB 031	110.6	1.6, 6.7
			AC 082	155.2	9.5
μ'	12 54.6	−57 11	017	34.9	4.0, 5.2

Open clusters

M/C	NGC	RA h m	Dec. ° '	Diameter min.	Mag.	No of stars
	4103	12 06.7	−61 15	7	7.4	45
	4349	12 24.5	−61 54	16	7.4	30
C98	4609	12 42.3	−62 58	5	6.9	40
C94	4755	12 53.6	−60 20	10	4.2	50+

(Jewel Box; cluster round κ Crucis)

Dark nebula

M/C	NGC	RA h m	Dec. ° '	Dimensions min.	Mag. of illum. star
C99		12 53	−63	400 × 300	26.2

(Coal Sack)

CENTAURUS

Open clusters

M/C	NGC	RA h m	Dec. ° '	Diameter min.	Mag.	No of stars
C97	3766	11 36.1	−61 37	12	5.3	100
C100	IC 2944	11 36.6	−63 02	15	4.5	Open cluster

(λ Centauri Cluster)

Carina and Volans are described with Map 22 and Vela with Map 21. Most of Centaurus is shown on Map 39, but the λ Centauri Cluster is just off that map.

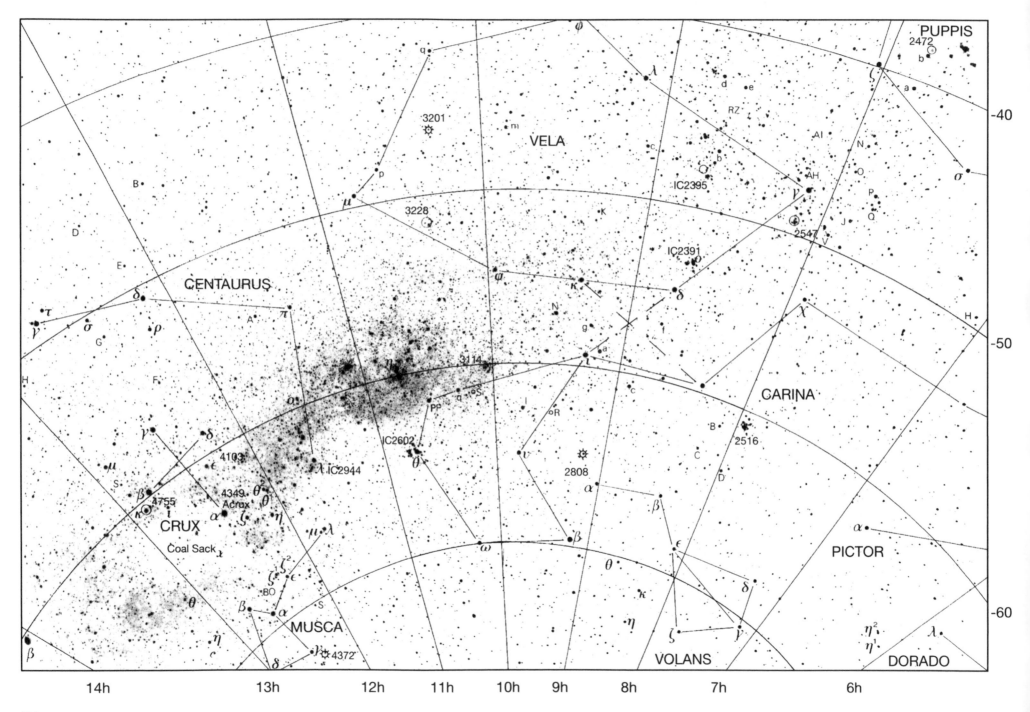

PUPPIS

2472
b

VELA

φ
q
λ
d
e
RZ
3201
m
AI
N
O
σ
p
IC2395
AH
γ
P
J
Q
μ
3228
K
2547
V

CENTAURUS

B
D
E
φ
δ
χ
τ
δ
π
κ
N
H
γ
σ
A
g
ρ
G
ι
a
CARINA
F
H
3114
c
o
η
S
B
γ
δ
q
PP
l
2516
ϵ
IC2602
$\circ$R
C
4103
λ IC2944
v
D
μ
θ
α
S
4349
θ^2
θ
2808
β
Acrux
θ
α
κ
4755
ι
α
ζ
η
β
PICTOR
CRUX
μ λ
ω
α
Coal Sack
θ
ϵ
δ
ζ^2
ζ^1
ϵ
κ
η^2
BO
λ
θ
η
η^1
β
S
ζ
γ
η
β
α
MUSCA
VOLANS
DORADO
η
γ
4372
c
δ

14h   13h   12h   11h   10h   9h   8h   7h   6h

-40

-50

-60

Map 24

URSA MAJOR

Virtually the whole of this map is occupied by Ursa Major, the third largest of all the constellations; it covers 1,280 square degrees, and is exceeded only by Hydra (1,303) and Virgo (1,294).

The seven leading stars of Ursa Major make up the pattern known variously as the Plough, King Charles' Wain and (in America) the Big Dipper. There are no stars of the first magnitude, but the pattern is so distinctive that it cannot be mistaken; from countries such as Britain it is circumpolar.

It is at once evident that Megrez is much the faintest of the seven Plough stars. Several of the astronomers of Classical times made it equal to the rest, but the evidence is very scanty, and Megrez does not seem to be a star of the type likely to show a permanent change in brightness.

Five of the Plough stars are travelling through space in the same direction at the same rate, and are roughly the same distance from us (around 65 light-years); they make up a true "moving cluster", and no doubt had a common origin. The remaining two, Alkaid and Dubhe, are moving in the opposite direction. The proper motions of stars are too slight to be noticed with the naked eye over periods of many centuries, but over many thousands of years the familiar Plough pattern will change

Merak and Dubhe are known as the Pointers, because they show the way to the Pole Star – which is well off the top part of the present map (it is shown in Maps D and 33). Unlike its companions, Dubhe is orange; the photograph shows it as yellow, but the colour is rather more pronounced than as shown here. It is a close binary; the companion is of magnitude 5, but as the separation is never as much as one second of arc this

rather a difficult pair. The revolution period is nearly 44.7 years.

The most celebrated star in Ursa Major is Zeta (Mizar). On any clear night its companion, 80 Ursæ Majoris or Alcor (4.0, A5) is clearly visible without optical aid – the classic example of a naked-eye pair. The separation is over 700 seconds of arc. Although Mizar and Alcor certainly make up a physically-connected system, they are at least a quarter of a light-year apart and probably rather more, so that their orbital period must amount to millions of years.

There is a minor mystery here. The Arabs of a thousand years ago stated categorically that Alcor was a difficult naked-eye object, and they even called it "Suha", the Forgotten One, because it could so easily be overlooked. In the thirteenth century a Persian writer, Al-Kazwini, even stated that "people test their eyesight by this star". Yet today it is obvious to anyone with normal sight. This is strange; the Arabs were noted for their keen eyes, and their skies were much darker than ours, because of the lack of light pollution.

It is not likely that there has been any real change. There has been a suggestion that "Suha" was not Alcor at all, but a much fainter star which telescopes show to lie between Mizar and Alcor. This star seems to have been discovered in 1681 by a German observer named Eimmart, using one of the small-aperture, long-focus refractors current at the time; it was re-discovered in 1723 by another German, whose name has not been preserved, and who seems to have had a limited knowledge of astronomy, since he believed that what he had seen was a new star or even a planet(!). He named it

Ludwig's Star, in honour of Ludwig V, Landgrave of Hesse. Ludwig's Star is well below naked-eye range, but if it were a few magnitudes brighter it would certainly be an optical test. If it ever shows signs of variability we would at least know what the Arabs meant – but it does not, and it lies well in the background beyond the Mizar group. A very small telescope will show it.

Mizar itself is an easy double, and was indeed the first pair to be detected telescopically -- in 1651 by the Jesuit astronomer Riccioli, best remembered today for drawing up a map of the Moon, and giving the lunar craters names which are still in use. The separation is over 14 seconds of arc. The magnitudes of the two components are respectively 2.3 and 4.0; both are of type A; the luminosities are 56 and 11 times that of the Sun (even Alcor could match more than ten Suns). The real separation is of the order of 400 astronomical units, so that here too the orbital period is very long.

Both components are spectroscopic binaries, and this is also true of Alcor, so that all in all the Mizar system consists of six suns. Indeed, Mizar A was the first spectroscopic binary to be detected – by the American astronomer EC Pickering, in 1889.

Xi Ursæ Majoris (3.8, G0) is an easy double with fairly equal components; it was the first binary to have its orbit computed (by Félix Savary, in 1828). Its neighbour Nu (3.5, K3) is also double; the separation is greater than with Xi, but the companion is only just above the tenth magnitude.

Below the centre of the photograph is a triangle made up of Psi (3.0, K1), Lambda (3.5, A2) and Mu (3.0, M0). We have here a good example of colour contrast;

Lambda is pure white while Mu is very red, so that the pair makes a fine sight through binoculars or a low-power telescope. Close to this triangle is an outwardly undistinguished star, 47 Ursæ Majoris (4.4, G0), which is important because in 1996 it was announced that slight movements indicated the presence of an orbiting planet. Whether this is true or not remains to be seen, and even if the planet exists it will be more like Jupiter than like the Earth, but certainly there is a chance that the star is the centre of a true planetary system.

The semi-regular variable Z Ursæ Majoris lies in the "bowl", not far from Delta. Its range is between magnitude 6.8 and 9.1, and there is a rough period of around 196 days. Since it never becomes really faint, and is easy to find, it is a favourite "training" object for new variable-star observers. The photograph shows it well, and, like all M-type stars, it is obviously red. Two more variables, R and VY, lie above and to the right of Dubhe. R is an ordinary Mira star; it can reach magnitude 6.7 when at its best, and when this photograph was taken it was almost at maximum, though the colour is not very pronounced. VY has an N-type spectrum, and is very red indeed; the magnitude fluctuates around 6, and there seems to be no definite period. It and R are fairly easy to locate because of the presence of a fairly distinctive curved line of faint stars nearby. The brightish star above and to the left is Lambda Draconis (4.0, M0); the photograph shows that it is considerably more highly-coloured than Dubhe.

In this region there are two very interesting galaxies, M81 and M82, close to 24 Ursæ Majoris (4.6, G2); a

Continued on page 116

URSA MAJOR

Brightest stars

Star		Proper name	Mag.	Spectrum	RA h m s	Dec. ° ' "
ε	Epsilon	Alioth	1.77v	A0	12 54 01.7	+55 57 35
α	Alpha	Dubhe	1.79	K0	11 03 43.6	+61 45 03
η	Eta	Alkaid	1.86	B3	13 47 32.3	+49 18 48
ζ	Zeta	Mizar	2.09	A2+A6	13 23 55.5	+54 55 31
β	Beta	Merak	2.37	A1	11 01 50.4	+56 22 56
γ	Gamma	Phad	2.44	A0	11 53 49.7	+53 41 41
ψ	Psi		3.01	K1	11 09 39.7	+44 29 54
μ	Mu	Tania Australis	3.05	M0	10 22 19.7	+41 29 58
ι	Iota	Talita	3.14	A7	08 59 12.4	+48 02 29
θ	Theta		3.17	F6	09 32 51.3	+51 40 38
δ	Delta	Megrez	3.31	A3	12 15 25.5	+57 01 57
o	Omicron	Muscida	3.36	G4	08 30 15.8	+60 43 05
λ	Lambda	Tania Borealis	3.45	A2	10 17 05.7	+42 54 52
ν	Nu	Alula Borealis	3.48	K3	11 18 28.7	+33 05 39
κ	Kappa	Al Kaprah	3.60	A0	09 03 37.5	+47 09 23
	h (23)		3.67	F0	09 31 31.7	+63 03 42
χ	Chi	Alkafzah	3.71	K0	11 46 03.0	+47 46 45
ξ	Xi	Alula Australis	3.79	G0	11 18 10.9	+31 31 45
υ	Upsilon		3.80v	F2	09 50 59.3	+59 02 19

Variable stars

	RA h m	Dec. ° '	Range (mags.)	Type	Period (d)	Spectrum
W	09 43.8	+55 57	7.9–8.6	W UMa	0.33	F+F
R	10 44.6	+68 47	6.7–13.4	Mira	301.7	M
TX	10 45.3	+45 34	7.1–8.8	Algol	3.06	B+F
VY	10 45.7	+67 25	5.9–6.5	Irreg.		N
VW	10 59.0	+69 59	6.8–7.7	Semi-reg.	125	M
ST	11 27.8	+45 11	7.7–9.5	Semi-reg.	81	M
CF	11 53.0	+37 43	8.5–12	Flare		B
Z	11 56.5	+57 52	6.8–9.1	Semi-reg.	196	M
RY	12 20.5	+61 19	6.7–8.5	Semi-reg.	311	M
T	12 36.4	+59 29	6.6–13.4	Mira	256.5	M
S	12 43.9	+61 06	7.0–12.4	Mira	225	S

Double stars

	RA h m	Dec. ° '	PA °	Sep. sec	Mags.	
σ²	09 10.4	+67 08	000	3.4	4.8, 8.2	Binary, 1067 y
α	11 03.7	+61 45	283	0.7	1.9, 4.8	Binary, 44.7 y
ν	11 18.5	+33 06	147	7.2	3.5, 9.9	
78	13 00.7	+56 22	057	1.5	5.0, 7.4	Binary, 116 y
ζ	13 23.9	+54 56	AB 152	14.4	2.3, 4.0	
			AC 071	708.7	2.1, 4.0	

Galaxies

M/C	NGC	RA h m	Dec. ° '	Mag.	Dimensions min.	Type
M81	3031	09 55.6	+69 04	6.9	25.7 × 14.1	Sb
(Bode's Nebula)						
M82	3034	09 55.8	+69 41	8.4	11.2 × 4.6	Pec.
	3077	10 03.3	+68 44	9.8	4.6 × 3.6	E2p
	3079	10 02.0	+55 41	10.6	7.6 × 1.7	Sb
	3184	10 18.3	+41 25	9.7	6.9 × 6.8	Sc
	3198	10 19.9	+45 33	10.4	8.3 × 3.7	Sc
	3310	10 38.7	+53 30	10.9	3.6 × 3.0	SBc
M108	3556	11 08.7	+55 57	10.0	8 × 1	Sc
M109	3992	11 55.0	+53 39	9.5	7 × 4	Sb
M101	5457	14 03.2	+54 21	7.7	26.9 × 26.3	Sc
(Pinwheel Galaxy)						
	5475	14 05.2	+55 45	12.4	2.2 × 0.6	Sa

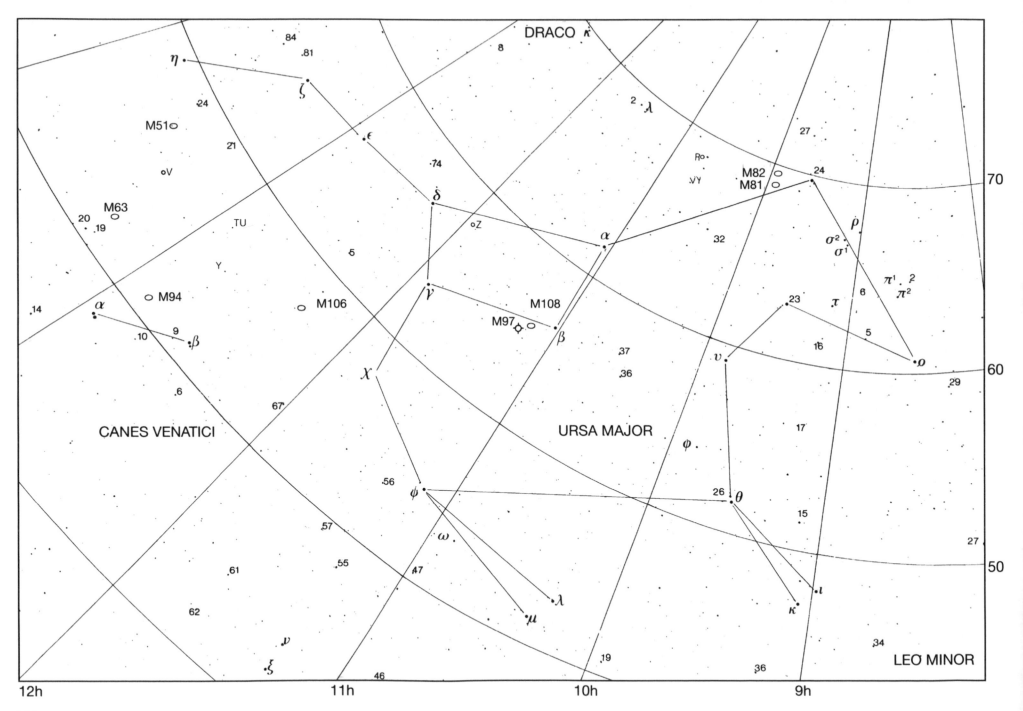

DRACO κ

η
ζ
84
.81
8

²λ
24
27

M51○
21
ε
.74
Ro○
M82 ○
M81 ○
24
70

δ
○Z
32
ρ
σ²
σ¹

M63○
M94○
TU
Y
ν
M106○
M108
M97◇○
β
23
π¹ 2
π²
6
τ
5

20
.19
α
.14
α
10
9
β
.6
67
.37
.36
ν
16
5
o
29
60

CANES VENATICI
χ
.56
URSA MAJOR
φ
17
27

57
ψ
26 θ
15

47
ω
55
61
λ
κ
ι
50

62
ν
μ
34

ξ
19
36
LEO MINOR

46

12h   11h   10h   9h

good "marker" is the little triangle consisting of Rho (4.8, M0), Sigma[1] (5.1, K5) and Sigma[2] (4.8, F7). Both galaxies were discovered by JE Bode in 1774. M81 is a fine spiral "only" a little over 8 million light-years away and therefore not too far beyond the Local Group; it is interacting with its neighbour M82, a very active system in which star formation is proceeding at an energetic rate. Both are easy telescopic objects, but they are not obvious on the map, though they are just visible. Photographs taken with adequate telescopes show a tremendous amount of structure, and the amateur will enjoy taking pictures of them – particularly if his telescope is equipped with a CCD! Another fine spiral in Ursa Major is M101, but this is just off the top of this photograph, above Mizar and Alkaid. It is shown on Map 34.

The edge-on spiral M108 lies close to Beta, the fainter of the Pointers; it is identifiable on the map without much difficulty. Adjoining it is the planetary nebula M97, which is much more elusive; it was discovered in 1781 by Pierre Méchain, Messier's friendly rival. It is nicknamed the Owl, because of the positioning of the two stars in it. It is about 10,000 light-years away, and can be rather difficult to identify, though it is just indicated on the photograph.

Canes Venatici, the Hunting Dogs, adjoins Ursa Major; it is described with Map 28. However, on this photograph there are indications of the Whirlpool Galaxy, M51, which is actually in Canes Venatici though the nearest bright star is Alkaid. The leader of the Dogs is Alpha[2] or Cor Caroli (2.9, A0 peculiar), seen to the left in this photograph.

Map 25

LYNX

This map shows almost the whole of Lynx, together with parts of Gemini, Hydra and Cancer. The cluster Præsepe is shown to the right of the centre of the photograph, both Castor and Pollux to the upper right, and the Sickle of Leo to the lower left.

Lynx is a dim constellation; indeed, it was once said that the name comes from the fact that an observer needs eyes like a lynx to see anything there at all! The brightest star is Alpha (3.1, M0) which is orange-red. No other stars in the constellation have been given Greek letters; next in order comes 38 (3.9, A3).

UX Lyncis is officially classed as a variable, but it has a very small range; it is, however, recognizable because of its colour. 38 is a multiple star. The globular cluster NGC 2419 (C25) is just identifiable on the photograph; it is interesting because it seems to be escaping from our Galaxy, in which case it will become what may be called an "intergalactic tramp".

The extreme northernmost part of Lynx is just over the upper border of this image, but is shown in the photograph with Map 11.

LYNX

Brightest stars

Star	Proper name	Mag.	Spectrum	RA h m s	Dec. ° ′ ″
α Alpha		3.13	M0	09 21 03.2	+34 23 33
38		3.92	A3	09 18 50.6	+36 48 09

Variable stars

	RA h m	Dec. ° ′	Range (mags.)	Type	Period (d)	Spectrum
RR	06 26.4	+56 17	5.6–6.0	Algol	9.95	A
R	07 01.3	+55 20	7.2–14.5	Mira	378.7	S
Y	07 28.2	+45 59	7.8–10.3	Semi-reg.	110	M

Double stars

	RA h m	Dec. ° ′	PA °	Sep. sec	Mags.
12	06 45.2	+59 27	AB 070	1.7	5.4, 6.0
			AC 308	8.7	7.3
			AD 256	170.0	10.6
19	07 22.9	+55 17	AB 315	14.8	5.6, 6.5
			AD 003	214.9	8.9
			BC 287	74.2	10.9
38	09 18.8	+36 48	AB 229	2.7	3.9, 6.6
			BC 212	87.7	10.8
			BD 256	177.9	10.7

Globular cluster

M/C	NGC	RA h m	Dec. ° ′	Diameter min.	Mag.
C25	2419	07 38.1	+38 53	4.1	10.4
(Intergalactic Tramp)					

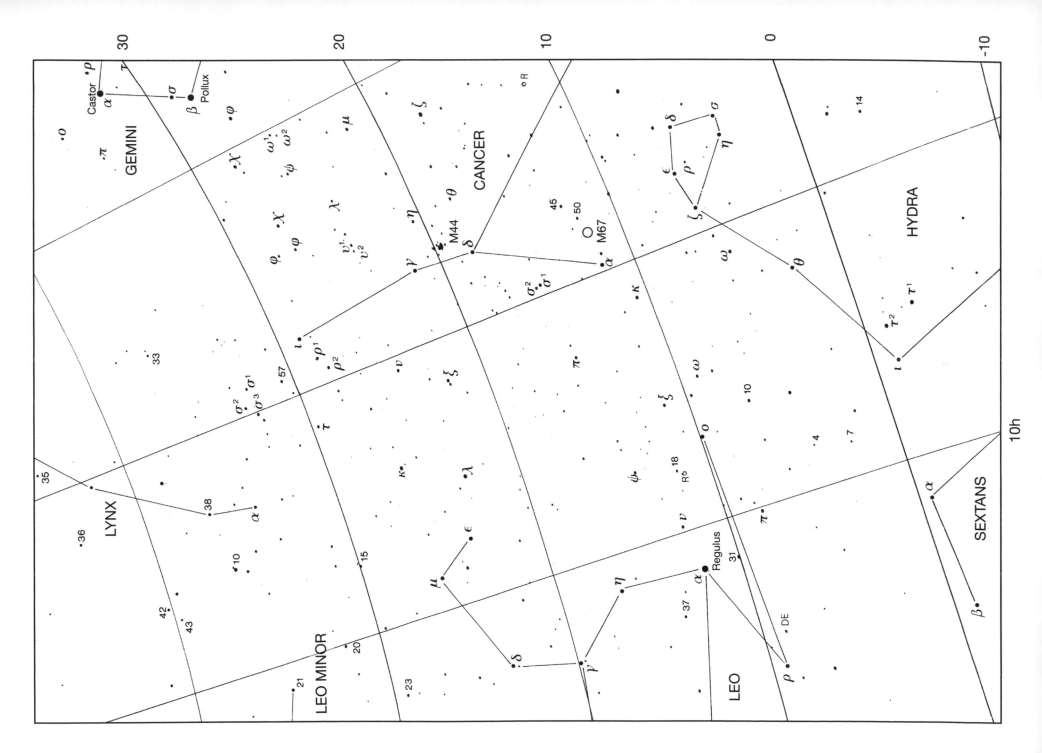

Map 26

CANCER

This map shows the whole of Cancer and Canis Minor and most of Sextans, together with part of Hydra and small portions of Leo and Gemini.

When the photograph was taken, Mars was on the borders of Gemini and Cancer, and of course dominates the scene; its magnitude was then above 0, and it far outshines the two first-magnitude stars on the map, Procyon in Canis Minor (lower right) and Regulus in Leo (upper left). Yet when Mars is at its faintest, it is barely brighter than Alphard in Hydra (lower left). Of course the ruddy hue is always very pronounced; it is interesting to compare Mars, as shown here, with bright red stars such as Antares and Betelgeux. The very name Antares means "The Rival of Mars" – the Greek equivalent of the war-god being Ares.

Leo appears here only in part, and at an unfamiliar angle; the Sickle extends to the upper right of Regulus. Below and to the right of Regulus, note the little triangle of stars; 18 Leonis (5.6, K5), 19 (6.4, F0) and the Mira-type variable R Leonis. Of these three, 18 is orange-yellow, 19 appears white and R, like all Mira stars, very red. At its peak R can rise to magnitude 4.4; in this photograph the magnitude was 7. To the left of Regulus are three stars with variable-star designations, DE, CX and TX, but in each case the range is less than a tenth of a magnitude; the median values are respectively magnitudes 5.6, 6.0 and 6.1.

Cancer, the Crab, is one of the dimmest Zodiacal constellations; in outline it is not unlike a very faint and ghostly Orion. The brightest star, Beta (3.6, K4) is decidedly orange. Above Beta on the photograph, and to the right, is the Mira variable R Cancri, which is here shown not far from its maximum magnitude of 6.1 – and, naturally, it is red. Even redder is X Cancri, a semi-regular variable with an N-type spectrum, clearly identifiable close to the little pair made up of Omicron[1] (5.2, A3) and Omicron[2] (4.8, A5). The magnitude of X at this time was 5.6, which is as bright as it ever becomes; it never drops below 7.5, so that it is always within binocular range, and it is one of the most highly-coloured stars in the entire sky.

Zeta Cancri, below and to the left of Mars in the photograph, is an interesting multiple system. The main component is a very close binary (AB); the separation is never greater than 1.2 seconds of arc. However, the third component (C) is an easy telescopic object; the separation is almost 5 seconds of arc, and the magnitudes of the components (AB+C) are 5.0 and 6.2. All three components are physically associated, and all three are yellowish. The distance from us is 52 light-years, so that on the scale the Zeta Cancri trio is reasonably close.

Cancer may be dim, but it is redeemed by the presence of two splendid open clusters. M44, known as the Beehive or the Manger although its official name is Præsepe, has been known since very early times, and was described by most of the astronomers of the Classical period, including Hipparchus and Ptolemy. Galileo, in 1610, described it as "not one star only, but a mass of more than forty small stars". It is an easy naked-eye object; on the photograph it looks decidedly bluish, but the leading stars are mainly white, yellowish or orange. Because it is large and scattered, it is probably best seen with binoculars or a wide-field telescope. It is flanked to either side by Delta Cancri (3.9, K0) and Gamma Cancri (4.7, AI). Because Præsepe is known as the Manger, Delta and Gamma have become the Asses – Asellus Australis and Asellus Borealis respectively.

Præsepe is a fairly old cluster, in age closer to the Hyades than to the Pleiades, but the second Messier object in Cancer, M67, has the distinction of being one of the very oldest known galactic clusters. In general, such clusters are disrupted by passing field stars, and lose their identity; but M67 lies almost 1,500 light-years above the crowded region of the galactic plane, so that it is moving in a relatively sparsely-populated area and is less subject to disruption. It is in the same binocular field as Alpha Cancri (4.2, A3) and is on the fringe of naked-eye visibility, but it does not show up well on the photograph. Not surprisingly, many of its leading stars have evolved off the Main Sequence, and are yellowish or orange in colour. Binoculars give excellent views of M67.

Canis Minor has already been described (Map 17). The Mira variable S Canis Minoris is easily seen here; note too the orange hue of Gamma (4.3, K3).

The head of Hydra is shown here; the brightest star in it is Zeta (3.1, K0). Alphard or Alpha Hydræ is prominent at the lower left of the map; it is of magnitude 2.0, and strongly orange. It is often nicknamed "the Solitary One". Most of Hydra is shown in Map 29, and is described there, together with Sextans.

CANCER

Brightest stars

Star	Proper name	Mag.	Spectrum	RA h	m	s	Dec. °	'	"
β Beta	Altarf	3.52	K4	08	16	30.9	+09	11	08
δ Delta	Asellus Australis	3.94	K0	08	44	41.0	+18	09	15

Variable stars

	RA h	m	Dec. °	'	Range (mags.)	Type	Period (d)	Spectrum
R	08	16.6	+11	44	6.1–11.8	Mira	361.6	M
V	08	21.7	+17	17	7.5–13.9	Mira	272.1	S
X	08	55.4	+17	14	5.6–7.5	Semi-reg.	195	N
T	08	56.7	+19	51	7.6–10.5	Semi-reg.	482	R–N
W	09	09.9	+25	15	7.4–14.4	Mira	393.2	M
RS	09	10.6	+30	58	6.2–7.7	Semi-reg.	120	M

Double star

	RA h	m	Dec. °	'	PA °	Sep. sec	Mags.	
ζ	08	12.2	+17	39	AB+C 088	5.7	5.0, 6.2	Binary, 1150 y
					AB 182	0.6	5.3, 6.0	Binary, 59.7 y
					AB+D 108	287.9	9.7	

Open clusters

M/C	NGC	RA h	m	Dec. °	'	Diameter min.	Mag.	No of stars
M44 (Præsepe)	2632	08	40.1	+19	59	95	3.1	50
M67	2682	08	50.4	+11	49	30	6.9	200

Galaxy

M/C	NGC	RA h	m	Dec. °	'	Mag.	Dimensions min.	Type
C48	2775	09	10.3	+07	02	10.3	4.5 × 3.5	Sa

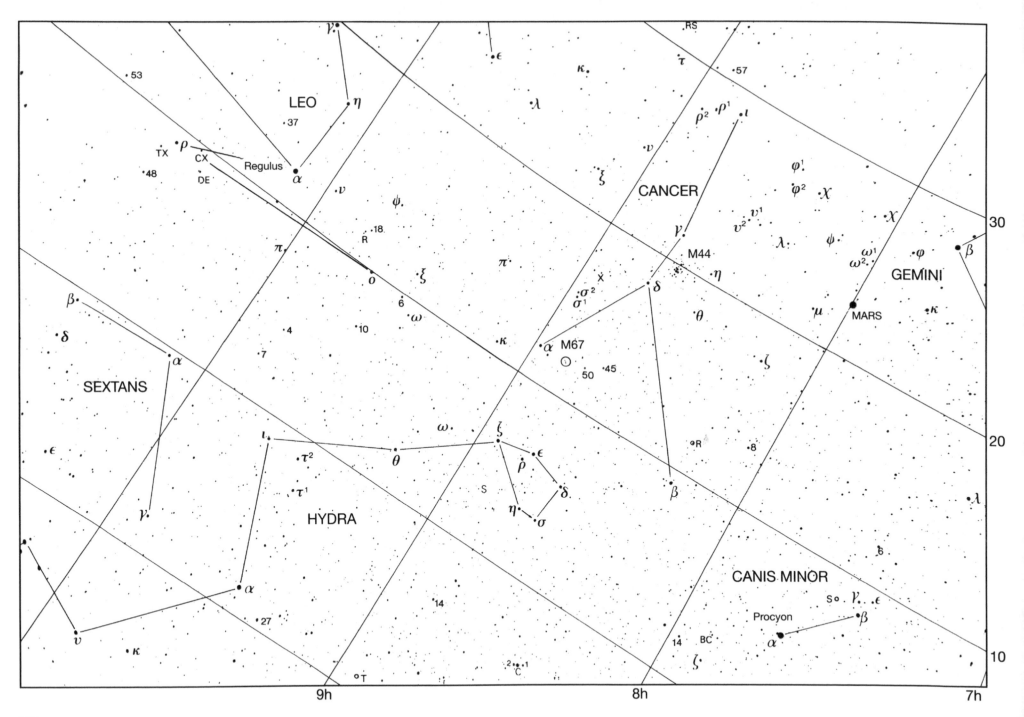

Map 27

LEO, LEO MINOR

This map contains the whole of Leo, one of the most imposing of the Zodiacal constellations. At the time of the photograph Jupiter was passing through it, and of course dominates the scene, since it is far brighter than any star; its magnitude is around –2.9, roughly one and a half magnitudes brighter than Sirius. Looking at this photograph, it is not easy to realize that the stars are suns, while Jupiter is simply a planet, shining by reflected sunlight and very small indeed compared with a normal star. Indeed the star close beside it in the photograph, Rho Leonis (3.8, B1) is 2,500 light-years away, and could match 16,000 Suns.

Leo is marked by the curved line of stars known as the Sickle; Regulus is the brightest member. (A good way to locate it is to use the Pointers in Ursa Major in the reverse direction, i.e., away from the Pole Star; sufficiently prolonged, the line will pass reasonably close to the Sickle.) Regulus is a pure white star, of type B7; it is 85 light-years away and 130 times as luminous as the Sun. The other stars in the Sickle are: Eta (3.5, A0); Gamma (2.0, K0+G7), Zeta (3.4, F0); Mu (3.9, K5); Epsilon (2.9, G0) and Lambda (4.3, K5). Mu and Lambda are slightly orange, but the most interesting star is Gamma, which is a particularly fine binary. The two components are rather unequal (magnitudes 2.2 and 3.5), but the separation is over 4 seconds of arc, so that a small telescope will show them adequately. The primary is orange; there is little colour in the secondary, but most observers describe it as slightly yellowish. It is worth noting that the radiant point of the Leonid meteors, seen annually around 17 November, lies close to Gamma.

The Mira variable R Leonis lies above and to the right of Regulus. At the time of this picture it was well away from maximum, and less bright than it appears in the photograph accompanying Map 26, but the colour can still be detected.

The other principal stars of Leo form a triangle, below and to the left of the Sickle; they are Beta or Denebola (2.1, A3), Delta (2.6, A4) and Theta (3.3, A2). Some of the astronomers of Classical times ranked Denebola as of the first magnitude, but today it is almost a magnitude fainter than Regulus. Any secular change seems to be most improbable; it is never wise to put too much faith in estimates of star brilliancies made centuries ago, and Denebola is a normal Main Sequence star, 39 light-years away and 17 times as powerful as the Sun.

Close to Regulus are CX Leonis (6.0, A0), DE (5.6, M3) and TX (5.7, A2). All three have variable star designations, but in each case the range is no more than a tenth of a magnitude. The red colour of DE can be seen in the photograph.

There are five Messier objects in Leo. M95 and M96 were discovered in 1781 by Pierre Méchain; M95 is a barred spiral and M96 a normal spiral, while the adjacent M105 is an elliptical system. They are fairly easy telescopic objects, but are not too easy to identify on the photograph, since they are rather small and have low surface brightness. M65 and M66, below Theta on the photograph, were also Méchain discoveries; both are spiral in form, and are just about within binocular range. They are 21 minutes of arc apart, and are genuinely associated; they are around 29 million light-years from us, and 180,000 light-years apart. The guide star for them is the rather orange 73 Leonis (5.5, K3).

(Like Messier, Pierre Méchain was a comet-hunter; he discovered eight in all. During his searches he often detected nebulous objects, and promptly passed the information on to Messier, with whom he was on excellent terms.)

Denebola makes up part of the boundary of the bowl of Virgo, which abounds in galaxies; this is described with Map 30. Adjoining it is Coma Berenices, also rich in galaxies; this too is contained in Map 28.

Parts of Canes Venatici, Ursa Major, Cancer, Sextans and Hydra appear on the present map; the famous open cluster M44 (Præsepe) is just included at the top right hand corner. Leo Minor, the junior Lion, contains little of note, though Beta is a close binary. One peculiar feature of Leo Minor is that Beta is the only star to have been allotted a Greek letter; the leader of the constellation is 46 Leonis Minoris (3.8, K0), which is orange.

LEO

Brightest stars

Star		Proper name	Mag.	Spectrum	RA h m s	Dec. ° ' "
α	Alpha	Regulus	1.35	B7	10 08 22.2	+11 58 02
γ	Gamma	Algieba	1.99	K0+G7	10 19 58.3	+19 50 30
β	Beta	Denebola	2.14	A3	11 49 03.5	+14 34 19
δ	Delta	Zosma	2.56	A4	11 14 06.4	+20 31 25
ε	Epsilon	Asad Australis	2.98	G0	09 45 51.0	+23 46 27
θ	Theta	Chort	3.34	A2	11 14 14.3	+15 25 46
ζ	Zeta	Adhafera	3.44	F0	10 16 41.4	+23 25 02
η	Eta		3.52	A0	10 07 19.9	+16 45 45
o	Omicron	Subra	3.52	A5	09 41 09.0	+09 53 32
ρ	Rho		3.85	B1	10 32 48.6	+09 18 24
μ	Mu	Rassalas	3.88	K2	09 52 45.8	+26 00 25
ι	Iota		3.94	F2	11 23 55.4	+10 31 45

Variable star

	RA h m	Dec. ° '	Range (mags.)	Type	Period (d)	Spectrum
R	09 47.6	+11 25	4.4–11.3	Mira	312.4	M

Double stars

	RA h m	Dec. ° '	PA °	Sep. sec	Mags.	
ω	09 28.5	+09 03	053	0.5	5.9, 6.5	Binary, 118 y
α	10 08.4	+11 58	307	176.9	1.4, 7.7	
γ	10 20.0	+19 51	AB 124	4.3	2.2, 3.5	Binary, 619 y
			AC 291	259.9	9.2	
			AD 302	333.0	9.6	
TX	10 35.0	+08 39	157	2.4	5.8, 8.5	
ι	11 23.9	+10 32	131	1.5	4.0, 6.7	Binary, 192 y
τ	11 27.9	+02 51	176	91.1	4.9, 8.0	

Galaxies

M/C	NGC	RA h m	Dec. ° '	Mag.	Dimensions min.	Type
	3190	10 18.1	+21 50	11.0	4.6 × 1.8	SG
M95	3351	10 44.0	+11 42	9.7	7.4 × 5.1	SBb
M96	3368	10 46.8	+11 49	9.2	7.1 × 5.1	Sb
	3377	10 47.7	+13 59	10.2	4.4 × 2.7	E5
M105	3379	10 47.8	+12 35	9.3	4.5 × 4.0	E1
	3384	10 48.3	+12 38	10.0	5.9 × 2.6	E7
	3412	10 50.9	+13 25	10.6	3.6 × 2.0	E5
	3489	11 00.3	+13 54	10.3	3.7 × 2.1	E6
	3521	11 05.8	−00 02	8.9	9.5 × 5.0	Sb
	3593	11 14.6	+12 49	11.0	5.8 × 2.5	Sb
	3596	11 15.1	+14 47	11.6	4.2 × 4.1	Sc
	3607	11 16.9	+18 03	10.0	3.7 × 3.2	E1
M65	3623	11 18.9	+13 05	9.3	10.0 × 3.3	Sb
C90	3626	11 20.1	+18 21	10.9	3.1 × 2.2	Sb
M66	3627	11 20.2	+12 59	9.0	8.7 × 4.4	Sb

LEO MINOR

Brightest star

Star	Proper name	Mag.	Spectrum	RA h m s	Dec. ° ' "
46	Præcipua	3.83	K0	10 53 18.6	+34 12 53

Variable stars

	RA h m	Dec. ° '	Range (mags.)	Type	Period (d)	Spectrum
R	09 45.6	+34 31	6.3–13.2	Mira	371.9	M
S	09 53.7	+34 55	7.9–14.3	Mira	233.8	M
RW	10 16.1	+30 34	6.9–10.1	Mira	?	N

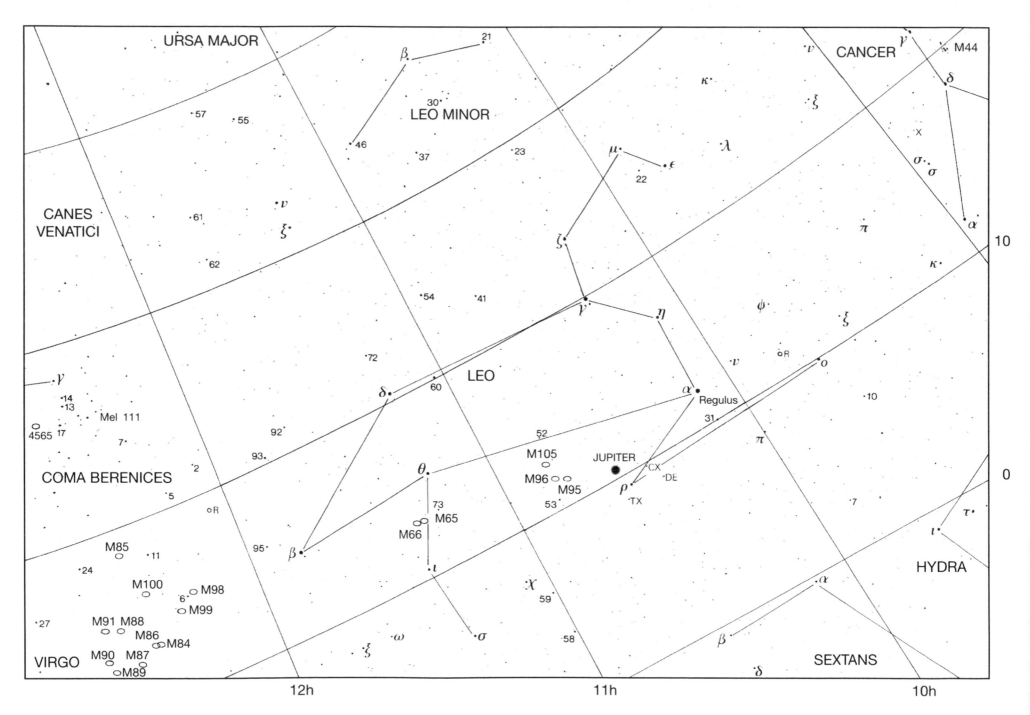

Map 28

CANES VENATICI, COMA BERENICES

Canes Venatici and Coma Berenices occupy much of this map.

Canes Venatici, the Hunting Dogs, adjoins Ursa Major; its leading star, Alpha[2] (2.9, A0 peculiar) is seen near the middle of the left-hand edge of the photograph. It has a romantic name – Cor Caroli, "Charles' Heart", bestowed on it in 1660 by Sir Charles Scarborough, physician to the English Court at the time of the restoration of King Charles II to the throne; it was said to shine with special brilliance on the night of the King's return, though no doubt this story is apocryphal! In fact the name honours Charles I, who had been so dramatically executed in 1649 at the end of the Civil War.

Cor Caroli is a fine double; rather confusingly the brighter component is listed as Alpha[2], while the secondary, of below magnitude 5, is Alpha[1]. The separation is over 19 seconds of arc, so that this is an extremely easy telescopic pair. Alpha[2] is interesting because its spectrum is variable, and it is known to have an unusually powerful magnetic field.

Forming a triangle with Alpha and Beta Canum Venaticorum (4.3, G0) is an exceptionally red semi-regular variable, Y, which was named "La Superba" by the nineteenth-century Italian astronomer Angelo Secchi because of its beauty. Its range is from magnitude 7.4 to 10; at the time of this photograph it was near maximum, so that the strong colour is very evident indeed.

There are several Messier objects in the Hunting Dogs. The most celebrated is M51, the Whirlpool, which lies at the edge of the constellation not far from Alkaid. This was the first spiral to have its form recognized, by Lord Rosse in 1845, using his home-made 72-inch (180-centimetre) reflector at Birr Castle in Ireland; it is face-on to us, and is spectacular when imaged with an adequate telescope, though it is barely visible in this photograph. Another spiral, M63, is not hard to locate because of its proximity to a distinctive little group of stars, of which the brightest is 20 Canum Venaticorum (4.7, F0). It is a "tighter" spiral than M51, and lies at a less favourable angle to us. M106, making a triangle with Psi and Chi Ursæ Majoris (3.4, A2) has an integrated magnitude of below 8, and on the photograph is only just traceable even though it has a longest diameter of as much as 18 seconds of arc. The other notable object in the constellation is the bright globular cluster M3, between Cor Caroli and Arcturus; it is just off the edge of this chart, but is shown in Map 34.

Coma Berenices abounds in galaxies, and there is also the large open cluster Mel 111, just below the orange star Gamma Comæ (4.3, K1). The Mira variable R Comæ can reach almost to magnitude 7 when at maximum, but at the time of this photograph was well below its best.

The brightest galaxy in the Coma cluster is M64, the so-called Black-Eye Galaxy. It is shown as a faint speck on the map; the best way to identify it is by the distinctive little group of fifth-magnitude stars (23, 26, 20 and 35 Comæ). It is within a degree of 35 (5.1, G8). All the other galaxies have integrated magnitudes of below 9, and are so crowded that they are not easy to pick out on a photograph taken to this scale. Of special interest is NGC 4565 (C38), which is the largest of the edgewise-on spiral galaxies, and through a modest telescope of, say, 10-centimetre (4-inch) aperture it appears as a bright, narrow streak.

CANES VENATICI

Brightest star

Star	Proper name	Mag.	Spectrum	RA h m s	Dec. ° ' "
α² Alpha²	Cor Caroli	2.90	A0p	12 56 01.6	+38 19 06

Variable stars

	RA h m	Dec. ° '	Range (mags.)	Type	Period (d)	Spectrum
Y	12 45.1	+45 26	7.4–10.0	Semi-reg.	157	N
(La Superba)						
U	12 47.3	+38 23	7.2–11.0	Mira	345.6	M
TU	12 54.9	+47 12	5.6–6.6	Semi-reg.	50	M
V	13 19.5	+45 32	6.5–8.6	Semi-reg.	192	M
R	13 49.0	+39 33	6.5–12.9	Mira	328.5	M

Double star

	RA h m	Dec. ° '	PA °	Sep. sec	Mags.
α²	12 56.0	+38 19	22.9	19.4	2.9, 5.5

Globular cluster

M/C	NGC	RA h m	Dec. ° '	Diameter min.	Mag.
M3	5272	13 42.2	+28 23	16.2	6.4

Galaxies

M/C	NGC	RA h m	Dec. ° '	Mag.	Dimensions min.	Type
	4151	12 10.5	+39 24	10.4	5.9 × 4.4	Pec. Strong UV source
	4214	12 15.6	+36 20	9.8	7.9 × 6.3	Irr.
M106	4258	12 19.0	+47 18	8.3	18.2 × 7.9	Sb
C21	4449	12 28.2	+44 06	9.4	5.1 × 3.7	Irr.
	4490	12 30.6	+41 38	9.8	5.9 × 3.1	Sc
C32	4631	12 42.1	+32 32	9.3	15.1 × 3.3	Sc
	4656–7	12 44.0	+32 10	10.4	13.8 × 3.3	Sc
M94	4736	12 50.9	+41 07	8.2	11.0 × 9.1	Sb
C29	5005	13 10.9	+37 03	9.8	5.4 × 2.7	Sb
	5033	13 13.4	+36 36	10.1	10.5 × 5.6	Sb
M51	5194	13 29.9	+47 12	8.4	11.0 × 7.8	Sc
(Whirlpool)						
M63	5055	13 15.8	+42 02	8.6	12.3 × 7.6	

COMA BERENICES

No star above magnitude 4

Variable stars

	RA h m	Dec. ° '	Range (mags.)	Type	Period (d)	Spectrum
R	12 04.0	+18 49	7.1–14.6	Mira	362.8	M
FS	13 06.4	+22 37	5.3–6.1	Semi-reg.	58	M

Galaxies

M/C	NGC	RA h m	Dec. ° '	Mag.	Dimensions min.	Type
M98	4192	12 13.8	+14 54	10.1	9.5 × 3.2	Sb
M99	4254	12 18.8	+14 25	9.8	5.4 × 4.8	Sc
M100	4321	12 22.9	+15 49	9.4	6.9 × 6.2	Sc
	4450	12 28.5	+17 05	10.1	4.8 × 3.5	Sb
M88	4501	12 32.0	+14 25	9.5	6.9 × 3.9	SBb
C35	4559	12 36.0	+27 58	9.8	10.5 × 4.9	Sc
C38	4565	12 36.3	+25 59	9.6	16.2 × 2.8	Sb
	4651	12 43.7	+16 24	10.7	3.8 × 2.7	Sop
	4725	12 50.4	+25 30	9.2	11.0 × 7.9	SBb
M64	4826	12 56.7	+21 41	8.5	9.3 × 5.4	Sb
(Black-Eye Galaxy)						

Open cluster

M/C	NGC	RA h m	Dec. ° '	Diameter min.	Mag.	No of stars
	Mel 111	12 25	+26	275	4.0	
(Coma Berenices)						

Globular cluster

M/C	NGC	RA h m	Dec. ° '	Diameter min.	Mag.
M53	5024	13 12.9	+18 10	12.6	7.7

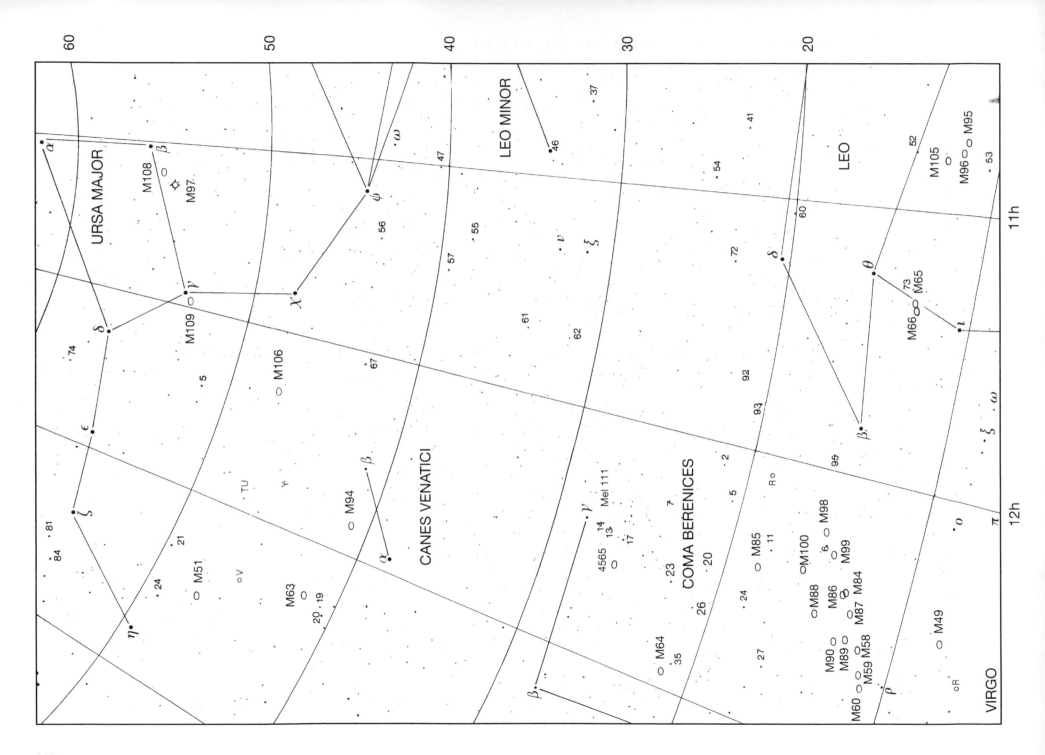

Map 29

HYDRA, SEXTANS

Here we have one of the more barren regions of the sky, occupied largely by Hydra, the immense Watersnake – sometimes identified with the multi-headed Lernæan hydra, one of Hercules' victims during the Twelve Labours. In fact, the only head of the celestial Hydra has also been shown in Map 26; it consists of Zeta (3.1, K0); Epsilon (3.4, G0); Delta 4.2, A0); Sigma (4.4, K2) and Eta (4.3, B3). Zeta is somewhat orange. Epsilon is a multiple system, but of the four main components two, A and B, are so close together that only a very large telescope will split them; the separation is at most 0.2 of a second of arc. This is a binary, with a very long period. The third member of the group, C, is easy; it is above the seventh magnitude, and almost 3 seconds of arc from the AB pair. Another star, D, shares in the same motion through space, and is presumably associated with the first three; it is a K0-type dwarf. Epsilon Hydræ is 110 light-years away; A and B combined have around 50 times the luminosity of the Sun. Theta (3.9, A0), not far from the Head, is an easy double.

The only star in this map as bright as the second magnitude is Alpha, or Alphard (2.0, K3); it is easy to see why it is nicknamed the Solitary One. It is clearly orange, and, as noted with Map 26, has been suspected of variability – though if there are any real fluctuations (which is dubious) the range must be no more than a few tenths of a magnitude. It has a distant ninth magnitude companion, but this is an optical pair, not a binary system.

S Hydræ, a Mira-type variable, makes up a triangle with Zeta and Eta, in the Watersnake's Head. It can rise to magnitude 7.4, but at the time of this photograph was well away from maximum, and is not too easy to make out. However, the semi-regular star U Hydræ, to the upper right of Nu (3.1, K2) is much easier. It has an N-type spectrum so that it is very red; its range is from magnitude 4.8 to 5.8 and is thus always well within binocular range, and it has a very long period (450 days, though as with all semi-regular variables the amplitude and the period vary somewhat from one cycle to another). In this photograph it is near maximum, and the colour is very evident indeed. It makes up a triangle with two fainter stars.

To the right-hand edge of the map, near the centre, is a little trio of stars (1, 2, C Hydræ). Below and to the right of them is the open cluster M48. Initially this was one of the "missing" objects in Messier's list – though Messier himself discovered it in 1771 – but it has been identified with NGC 2548. It lies on the borders of Hydra and Monoceros, and in fact the nearest reasonably bright star is Zeta Monocerotis (4.3, G2). It is not prominent, and there are various other star-groups nearby which make it rather hard to identify, but its three main stars form an equilateral triangle, which tends to mark it out.

Sextans has no star brighter than Alpha (4.5, B5). Not far from it is the Spindle Galaxy, NGC 3115 (C53), which is of just below the ninth magnitude; it is just detectable on the photograph, but telescopically it tends to be a somewhat elusive object. Gamma Sextantis is a close binary. Sextans was one of the constellations added to the sky by Hevelius; previously it had been included in Hydra.

HYDRA

Brightest stars

Star	Proper name	Mag.	Spectrum	RA h m s	Dec. ° ' "
α Alpha	Alphard	1.98	K3	09 27 35.2	−08 39 31
γ Gamma		3.00	G5	13 18 55.2	−23 10 17
ζ Zeta		3.11	K0	08 55 23.6	+05 56 44
ν Nu		3.11	K2	10 49 37.4	−16 11 37
π Pi		3.27	K2	14 06 22.2	−26 40 56
ε Epsilon		3.38	G0	08 46 46.5	+06 25 07
ξ Xi		3.54	G7	11 33 00.1	−31 51 27
λ Lambda		3.61	K0	10 10 35.2	−12 21 15
μ Mu		3.81	K4	10 26 05.3	−16 50 11
θ Theta		3.88	A0	09 14 21.8	+02 18 51

Variable stars

	RA h m	Dec. ° '	Range (mags.)	Type	Period (d)	Spectrum
RT	08 29.7	−06 19	7.0–11.0	Semi-reg.	253	M
S	08 53.6	+03 04	7.4–13.3	Mira	256.4	M
T	08 55.7	−09 08	6.7–13.2	Mira	289.2	M
X	09 35.5	−14 42	8.0–13.6	Mira	301.4	M
U	10 37.6	−13 23	4.8–5.8	Semi-reg.	450	N
V	10 49.2	−20 59	6.0–12.5	Mira	533	N+E
R	13 29.7	−23 17	4.0–10.0	Mira	389.6	M
TT	11 13.2	−26 28	7.5–9.5	Algol	6.95	A+G
HZ	11 26.3	−25 45	7.6–8.2	Semi-reg.	95	M

Double stars

	RA h m	Dec. ° '	PA °	Sep. sec	Mags.	
ε	08 46.8	+06 25	AB 295	0.2	3.8, 4.7	Binary, 890 y
			AB+C 281	2.8	6.8	
θ	09 14.4	+02 19	197	29.4	3.9, 9.9	
α	09 17.6	−08 40	153	283.1	2.0, 9.5	
β	11 52.9	−33 54	008	0.9	4.7, 5.5	

Open cluster

M/C	NGC	RA h m	Dec. ° '	Diameter min.	Mag.	No of stars
M48	2548	08 13.8	−05 48	54	5.8	80

Globular clusters

M/C	NGC	RA h m	Dec. ° '	Diameter min.	Mag.
M68	4590	12 39.5	−26 45	12.0	8.2
C66	5694	14 39.6	−26 32	3.6	10.2

Galaxies

M/C	NGC	RA h m	Dec. ° '	Mag.	Dimensions min.	Type
	3585	11 13.3	−26 45	10.0	2.9 × 1.6	E5
	3621	11 18.3	−32 49	9.9	10.0 × 6.5	Sc

SEXTANS

No star above magnitude 4

Galaxies

M/C	NGC	RA h m	Dec. ° '	Mag.	Dimensions min.	Type
C53	3115	10 05.2	−07 43	9.1	8.3 × 3.2	E6
(Spindle Galaxy)						
	3166	10 13.8	+03 26	10.6	5.2 × 2.7	SBa
	3169	10 14.2	+03 28	10.4	4.8 × 3.2	Sb

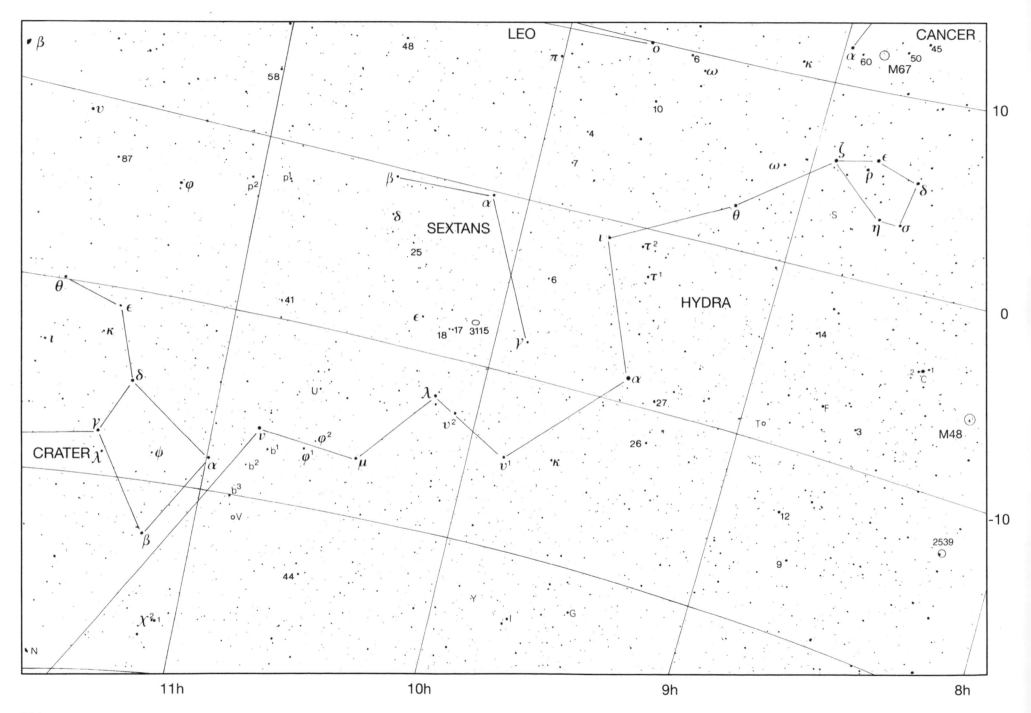

Map 30

VIRGO

Most of Virgo is contained in this map, together with parts of Coma, Crater, Corvus, Hydra, Libra and Boötes. The only first-magnitude star is Spica, but there are many interesting objects – notably in the "bowl" of Virgo, bounded on the far side by Beta Leonis or Denebola (see Map 27). The "bowl" is crowded with galaxies. On the scale of this photograph they are not easy to identify, but sweeping the area with a rich-field telescope is very rewarding.

Spica is 260 light-years away. It is an excessively close binary; the real distance between the two components must be less than 21 million kilometres (13 million miles), and the orbital period is a mere 4 days, so that normal telescopes show Spica as a single star. The total luminosity is over 2,000 times that of the Sun; both components are of spectral type B.

Gamma Virginis, at the base of the "bowl", is a celebrated binary. (It is distinguished by several proper names – Arich, Porrima and Postvarta – which are sometimes used even today.) The two components are virtual twins. Each is of magnitude 3.5, and of type F. The orbital period is just over 171 years, so that the position angle and separation change markedly over a few decades. In 1920 the separation was over 6 seconds of arc; it is now down to less than 3 seconds, and by the year 2010 will have decreased to below one second. The real separation between the twins ranges between 3 and 70 astronomical units. The components have been described as slightly yellowish, but few observers will see any colour in them. The distance from us is 36 light-years; each star is 3½ times as luminous as the Sun.

Of the other stars in the "bowl", Delta (3.4, M3) is a beautiful red giant, 120 times as powerful as the Sun and 147 light-years away. Epsilon (2.8, G9) is equal to 60 Suns. The other two members of the pattern are Beta (3.6, F8) and Eta (3.9, A2). Both have been suspected of variability, but the evidence is very slender indeed.

The Mira variable R Virginis forms a triangle with Delta and Epsilon. At the time of this photograph it was below the seventh magnitude, and the colour is not noticeable, but at its best it can rise to magnitude 6. SS Virginis, above and to the left of Eta, is also a Mira star, but has a smaller range (6.0 to 9.6) and is of type N, so that it is intensely red; the colour comes out well in the photograph. A third Mira star, S Virginis, lies between Spica and Zeta (3.4, A3); it is close to 74 Virginis, which is of type M3 and obviously red. The official magnitude of 74 Virginis is 4.7, but it is listed as a suspected variable, and in this photograph it is not much fainter than Theta (4.4, A1), so that the suspicions may well be justified. ET Virginis (4.9, M2) is also red; it appears below and to the right of Lambda (4.5, A0). It has a variable-star designation, but the range is only about a tenth of a magnitude.

Sigma Virginis (4.8, M2) is another red star, making up a triangle with Delta and Epsilon. To its left on the map is CW Virginis, which is a magnetic variable but almost constant in light; its magnitude is 4.9. The little pattern of stars below Sigma in the photograph is a help in identification.

Theta (4.4, A1), Tau (4.6, A3) and Phi (4.8, F8) have faint companions, but are not particularly notable. Of greater immediate interest is a single star, 70 Virginis, above and to the left of Epsilon. Its magnitude is 5.0, and it is a G5 star slightly less luminous than the Sun according to the Cambridge catalogue; it is a near neighbour, at only 33 light-years. This is one of the stars suspected of being orbited by a massive planet, though as yet the evidence is far from complete.

M104, below Chi (4.7, K2) and the red Psi (4.8, M3) on the map, is the celebrated Sombrero Galaxy, discovered by Méchain in 1781. It is almost edge-on to us, and earns its nickname because of the dark dust-lane running across it and giving it an aspect which is said to resemble a Mexican hat! Telescopically it is not a difficult object, and it bears high magnification well. The lines of stars below it in the photograph will be helpful in locating it.

Of the galaxies in the "bowl", the brightest is the giant elliptical M87, with the famous "jet" and probably a black hole in its centre. The Virgo cluster contains systems of all types, and is far more populous than our own sparse Local Group.

The globular cluster NGC 5634, between Iota and Mu, is just visible in the photograph, but it is very inconspicuous, and can be elusive telescopically.

Libra adjoins Virgo; only part of it appears here, but this includes Alpha, which is a wide double easily split with binoculars; it is described more fully in Map 37. Part of Corvus is shown; note the neat little pair of Delta (2.9, B5) and Eta (4.3, F0). The two are not genuinely associated; Eta is 29 light-years away, while the distance of Delta is 35 light-years.

The red variable R Hydræ is just on the photograph, at the very bottom to the left of Gamma Hydræ (3.0, G5). At its best it may rise to magnitude 4, so that it is then an easy naked-eye object. Its most interesting feature is that its period has shortened in recent times; in 1800 it was about 500 days, but is now less than 390 days. There seems no doubt that we are watching a real change in the star's evolution. The minimum magnitude is no fainter than 10.

VIRGO

Brightest stars

Star		Proper name	Mag.	Spectrum	RA h m s	Dec. ° ' "
α	Alpha	Spica	0.98v	B1	13 25 11.5	−11 09 41
γ	Gamma	Arich	2.75	F0+F0	12 41 39.5	−01 26 57
ε	Epsilon	Vindemiatrix	2.83	G9	13 02 10.5	+10 57 33
ζ	Zeta	Heze	3.37	A3	13 34 41.5	−00 35 46
δ	Delta	Minelauva	3.38	M3	12 55 36.1	+03 23 51
β	Beta	Zavijava	3.61	F8	11 50 41.6	+01 45 53
	109		3.72	A0	14 46 14.9	+01 53 34
μ	Mu	Rijl al Awwa	3.88	F3	14 43 03.5	−05 39 30
η	Eta	Zaniah	3.89	A2	12 19 54.3	−00 40 00

Variable stars

	RA h m	Dec. ° '	Range (mags.)	Type	Period (d)	Spectrum
X	12 01.9	+09 04	7.3–11.2	?		F
SS	12 25.3	+00 48	6.0–9.6	Mira	354.7	N
R	12 38.5	+06 59	6.0–12.1	Mira	145.6	M
U	12 51.1	+05 33	7.5–13.5	Mira	206.8	M
S	13 33.0	−07 12	6.3–13.2	Mira	377.4	M
RS	14 27.3	+04 41	7.0–14.4	Mira	352.8	M

Double stars

	RA h m	Dec. ° '	PA °	Sep. sec	Mags.	
17	12 22.5	+05 18	337	20.0	6.6, 9.4	
γ	12 41.7	−01 27	287	3.0	3.5, 3.5	Binary, 171.4 y
θ	13 09.9	−05 32	343	7.1	4.4, 9.4	
84	13 43.1	+03 32	229	2.9	5.5, 7.9	
τ	14 01.6	+01 33	290	80.0	4.3, 9.6	
φ	14 28.2	−02 14	110	4.8	4.8, 9.3	

Galaxies

M/C	NGC	RA h m	Dec. ° '	Mag.	Dimensions min.	Type
M61	4303	12 21.9	+04 28	9.7	6.0 × 5.5	Sc
M84	4374	12 25.1	+12 53	9.3	5.0 × 4.4	E1
M86	4406	12 26.2	+12 57	9.2	7.4 × 5.5	E3
M49	4472	12 29.8	+08 00	8.4	8.9 × 7.4	E4
M87	4486	12 30.8	+12 24	8.6	7.2 × 6.8	E1 Virgo A
	4699	12 49.0	−08 40	9.6	3.5 × 2.7	Sa
	4535	12 34.3	+08 12	9.8	6.8 × 5.0	SBc
M89	4552	12 35.7	+12 33	9.8	4.2 × 4.2	E0
M90	4569	12 36.8	+13 10	9.5	9.5 × 4.7	Sb
M58	4579	12 37.7	+11 49	9.8	5.4 × 4.4	Sb
M104 (Sombrero)	4594	12 40.0	−11 37	8.3	8.9 × 4.1	Sb
M59	4621	12 42.0	+11 39	9.8	5.1 × 3.4	E3
M60	4649	12 43.7	+11 33	8.8	7.2 × 6.2	E1
	4636	12 42.8	+02 41	9.6	6.2 × 5.0	E1
C52	4697	12 48.6	−05 48	9.3	6.0 × 3.8	E4
	4699	12 49.0	−08 40	9.6	3.5 × 2.7	Sa
	4753	12 52.4	−01 12	9.9	5.4 × 2.9	Pec.

HYDRA

Variable star

	RA h m	Dec. ° '	Range (mags.)	Type	Period (d)	Spectrum
R	13 29.7	−23 17	4.0–10.0	Mira	390	M

Almost the whole of Hydra is shown on Map 29, and objects are listed there, but R is just off that map.

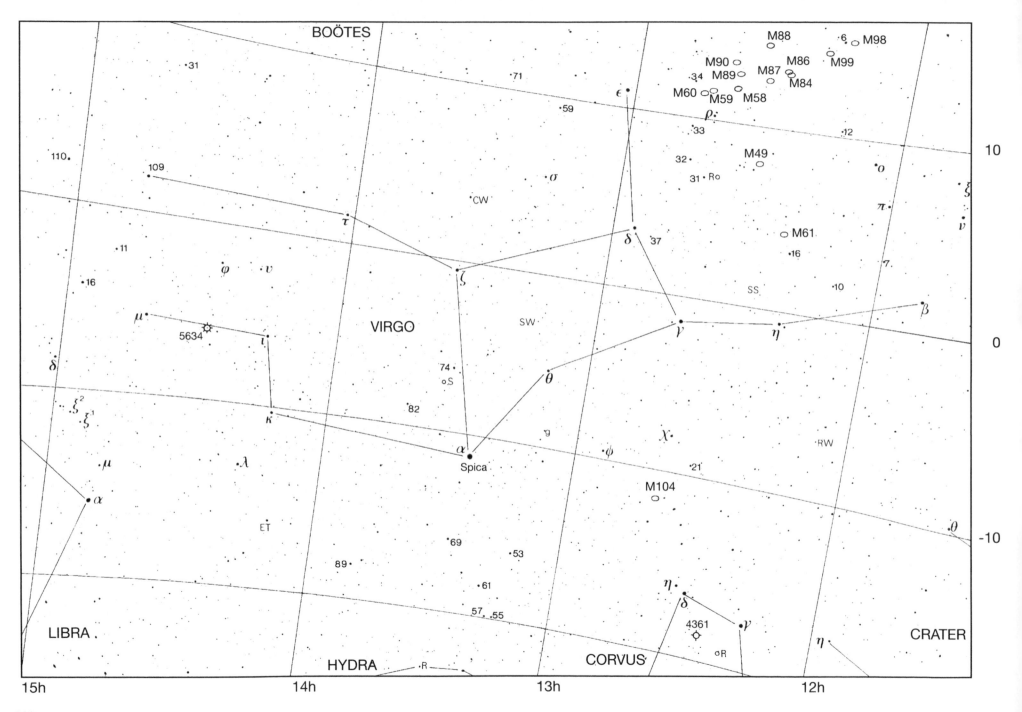

BOÖTES

M88 6 M98

31

71 M90 M87 M86 M99
M89 M84
34
59 M60 M58
ε M59

110 ρ

109 33 12

σ 32 M49 o

31 Ro π

CW δ 37 M61 7

11 16

φ υ ζ SS 10

16 SW β

μ VIRGO γ η

5634 ι 74 θ

δ oS

ξ² 82 χ RW
ξ¹ κ 9

α φ

μ λ Spica 21

M104

α

ET

69

89 53

η

61 δ

57 55 4361 γ

LIBRA η
CRATER

HYDRA CORVUS
R oR

15h 14h 13h 12h

10

0

-10

ξ
υ

θ

Map 31

CORVUS, CRATER

This map covers a rather barren region of the sky. Most of it is occupied by the central part of Hydra, together with Crater, Corvus, Sextans, and small sections of Virgo, Leo and Antlia. Alphard, the only bright star in Hydra, lies to the extreme right of the map; its orange hue is very obvious. Nu Hydræ (3.1, K2) appears slightly above the centre of the photograph; it too is clear orange. Mu Hydræ (3.8, K4), to the right and rather above Nu, has a similar spectrum, but its colour is much less pronounced. Not all K-stars are alike!

Three variables in Hydra are worth noting; two of them (V and FF) did not show up in the photograph together with Map 29 because they were then too near their minima, but they are detectable in the present photograph. U Hydræ is of type N, and extremely red; it is semi-regular, with a range of from magnitude 4.8 to 5.8. It lies above and to the right of Nu. Immediately above it is a little trio of stars, together with the Mira variable FF Hydræ, which is just identifiable here. Below Nu, and below and to the right of Alpha Crateris, is a triangle of faint stars of which the most prominent is b³ (5.3, F6). Just below this triangle is V Hydræ, again with an N-type spectrum; although at the time of this photograph it was well away from maximum (magnitude 6) its deep crimson colour is still striking. It has the unusually long period of 533 days.

Also in Hydra is the globular cluster M68, below Beta Corvi and in line with Delta and Beta; it lies within a degree of a fifth-magnitude double star, and is just identifiable on the photograph. Telescopically it shows up as a perfectly round patch; its real distance from us is about 39,000 light-years. It was discovered by Méchain in 1780.

Corvus contains no bright stars, but its leaders make up a quadrilateral which is very easy to identify, particularly in view of the fact that the surrounding areas are so barren. The main stars are Gamma (2.6, B8); Beta (2.7, G5); Delta (2.9, B9) and Epsilon (3.0, K2); the orange hue of Epsilon is detectable. Delta makes up a wide pair with Eta (4.3, F0). All the four leaders of Corvus have been suspected of variability, but with no firm evidence. Inside the quadrilateral, Zeta (5.2, B8) makes up a very wide pair with a fainter star. The planetary nebula NGC 4361 is just identifiable; so too is the red Mira variable R Corvi, though it was well away from maximum at the time of the photograph. Outside the quadrilateral, note the orange tint of 6 Corvi (5.7, K1).

The celestial Crow also contains the celebrated interacting galaxies nicknamed the Antennæ (NGC 4038–9; C60–1). Unfortunately they are below the eleventh magnitude, and therefore too faint to be shown here.

Crater, the Cup – in mythology, said to be the goblet of the wine god Bacchus – is very obscure; its main stars, Alpha (4.1, K0), Gamma (4.1, A5) and Delta (3.6, G8) form a triangle to the left of Nu Hydræ. Gamma has a faint companion at a separation of over 5 seconds of arc. The very red variable R Crateris lies in the same low-power telescopic field with Alpha, but never exceeds the eighth magnitude and is not shown.

The Sombrero Galaxy, M104 in Virgo, appears on this map, though it is easier to identify in the photograph with Map 30. The whole of Sextans is included, Gamma is a close binary; the Spindle Galaxy, NGC 3115 (C53) is hard to identify here.

Antlia is a very dim constellation, with a dubious claim to separate identity. Its brightest star, the orange Alpha (4.2, M0) lies near the bottom of the photograph, and makes up a pair with its neighbour Delta (5.6, A0).

CORVUS

Brightest stars

Star		Proper name	Mag.	Spectrum	RA h m s			Dec. ° ' "		
γ	Gamma	Minkar	2.59	B8	12 15 48.3			−17 32 31		
β	Beta	Kraz	2.65	G5	12 34 23.2			−23 23 48		
δ	Delta	Algorel	2.95	B9	12 29 51.8			−16 30 55		
ε	Epsilon		3.00	K2	12 10 07.4			−22 37 11		

Variable stars

	RA h m	Dec. ° '	Range (mags.)	Type	Period (d)	Spectrum
R	12 19.6	−19 15	6.7–14.4	Mira	317	M
SV	12 49.8	−15 05	6.8–7.6	Semi-reg.	70	M

Double star

	RA h m	Dec. ° '	PA °	Sep. sec	Mags.
δ	12 29.9	−16 31	214	24.2	3.0, 9.2

Planetary nebula

M/C	NGC	RA h m	Dec. ° '	Dimensions sec.	Mag.	Mag. of central star
	4361	12 24.5	−18 48	45 × 110	10.3	13.2

Galaxies

M/C	NGC	RA h m	Dec. ° '	Mag.	Dimensions min.	Type
C60	4038	12 01.9	−18 52	11.3	2.6 × 1.8	Sc Antennæ
C61	4039	12 01.9	−18 53	13	3.2 × 2.2	Smp Antennæ

CRATER

Brightest star

Star		Proper name	Mag.	Spectrum	RA h m s			Dec. ° ' "		
δ	Delta		3.56	G8	11 19 20.4			−14 46 43		

Double star

	RA h m	Dec. ° '	PA °	Sep. sec	Mags.
γ	11 24.9	−17 41	096	5.2	4.1, 9.6

HYDRA

Globular cluster

M	NGC	RA h m	Dec. ° '	Mag.	Dim.	Type
M68	4590	12 39.5	−26 45	8.2	12.0	

Almost the whole of Hydra is shown on Map 29, and objects are listed there, but M68 is just off that map.

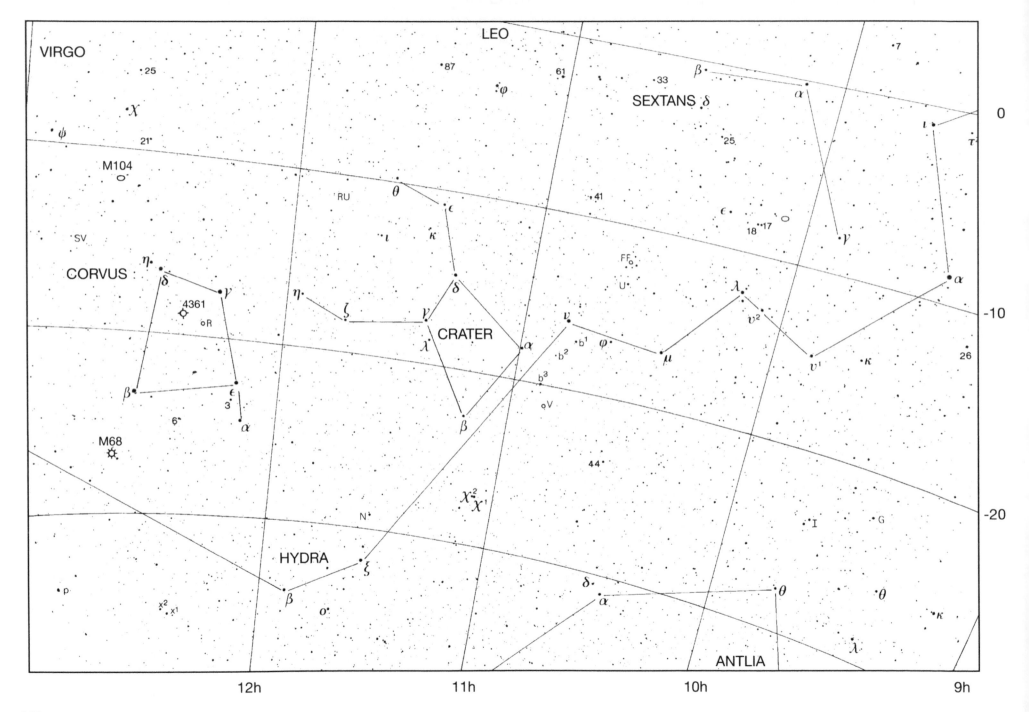

Map 32

TRIANGULUM AUSTRALE, APUS, MUSCA AUSTRALIS, CHAMÆLEON

This map shows the whole of Crux, Musca and Apus, with Chamæleon and Volans, most of Triangulum Australe and Carina, and sections of Octans, Circinus, Centaurus, Vela and Pictor. The Eta Carinæ nebula shows up beautifully. The Pointers, Alpha and Beta Centauri, appear to the left-hand edge of the photograph. The False Cross is seen to the right, and it is interesting to compare it with the Southern Cross.

The three main stars of Triangulum Australe really do form a triangle; they are Alpha (1.9, K2), Beta (2.8, F5) and Gamma (2.9, A0). Unfortunately Beta is just off the edge of this map – it is shown on Map 39. Alpha is decidedly reddish; it is 55 light-years away, and 96 times as luminous as the Sun.

X Trianguli Australe, rather below right of Gamma, is a very red N-type irregular variable; at the time when the photograph was taken it was near its maximum of magnitude 8. R, a Cepheid variable, can be made out. T, again near Gamma, was once regarded as variable, but is now thought to be constant at magnitude 6.8.

Apus (formerly Avis Indica, the Bird of Paradise) has no bright stars, but all its leaders are reddish: Alpha (3.8, K5), Gamma (3.9, K0), Beta (4.2, K0), Delta (4.7, M4) and Zeta 4.8, K5) – this makes the constellation quite distinctive. Delta is made up of two widely-separated components – they are over 100 seconds of arc apart, but seem to be physically connected; the second star is of type K5. Theta, detectable on the map, is a red semi-regular variable. The globular cluster NGC 6101

(C107) can be made out in the guise of a faint star.

Also shown is R Apodis, which has a variable star designation but is now thought to be constant at magnitude 5.3. If it does fluctuate, it has a very small range.

Chamæleon has no star brighter than Alpha (4.1, F6).

Musca Australis, the Southern Fly, is now generally known simply as Musca. (There was once a Northern Fly, Musca Borealis, created by J E Bode, but this has long since been swallowed up by Aries.) The Southern Fly is a compact little constellation; its brightest stars are Alpha (2.7, B3) and Beta (3.0, B3), shown at the bottom left of the photograph. Slightly above and to the right of these two is Epsilon (4.1, M5), which is orange-red. Still further above and to the right is a neat little pair with good colour contrast, made up of the white Lambda (3.6, A5) and the orange Mu (4.7, M2). In the photograph the colour contrast is unmistakable, and the pair is well worth looking at with binoculars or a telescope. Beta Muscæ is double, but the separation is always below one and a half seconds of arc; Alpha is a much wider pair, but the secondary component is very faint. Theta, to the left of the main pattern of Musca, is an easy telescopic double, though there is nothing special about it.

Two variables in Musca appear in the photograph. R, just below Alpha, is a Cepheid; the magnitude here was about 6.7, and the star is not too easy to identify. BO Muscæ, above Beta, is more prominent; the magnitude is about 6, and the spectral type is M, so that the star is

decidedly red. It seems to be irregularly variable.

Musca contains two fairly bright globular clusters, NGC 4833 (C105), near Delta, and NGC 4372 (C108), just below and to the right of Gamma on the map. Both are on the fringe of visibility with good binoculars, and are easy to see telescopically.

TRIANGULUM AUSTRALE

Brightest stars

Star	Proper name	Mag.	Spectrum	RA h m s	Dec. ° ' "
α Alpha	Atria	1.92	K2	16 48 39.8	−69 01 39
β Beta		2.85	F5	15 55 08.4	−63 25 50
γ Gamma		2.89	A0	15 18 54.5	−68 40 46
δ Delta		3.85	G2	16 15 26.2	−63 41 08

Variable stars

	RA h m	Dec. ° '	Range (mags.)	Type	Period (d)	Spectrum
X	15 14.3	−70 05	8.1–9.1	Irreg.		N
R	15 19.8	−66 30	6.4–6.9	Cepheid	3.39	F–G
S	16 01.2	−63 47	6.1–6.8	Cepheid	6.32	F

Open cluster

M/C	NGC	RA h m	Dec. ° '	Diameter min.	Mag.	No of stars
C95	6025	16 03.7	−60 30	12	5.1	60

APUS

Brightest stars

Star	Proper name	Mag.	Spectrum	RA h m s	Dec. ° ' "
α Alpha		3.83	K5	14 47 51.6	−79 02 41
γ Gamma		3.89	K0	16 33 27.1	−78 53 49

Variable stars

	RA h m	Dec. ° '	Range (mags.)	Type	Period (d)	Spectrum
θ	14 05.3	−76 48	6.4–8.6	Semi-reg.	119	M
S	15 04.3	−71 53	9.5–15	R CrB		R

(R Apodis, magnitude 5.3, has been suspected of variability.)

Double star

	RA h m	Dec. ° '	PA °	Sep. sec	Mags.
δ	16 20.3	−78 42	012	102.9	4.7, 5.1

Globular cluster

M/C	NGC	RA h m	Dec. ° '	Diameter min.	Mag.
	IC 4499	15 00.3	−82 13	7.6	10.6
C107	6101	16 25.8	−72 12	10.7	9.3

MUSCA AUSTRALIS

Brightest stars

Star	Proper name	Mag.	Spectrum	RA h m s	Dec. ° ' "
α Alpha		2.69v	B3	12 37 11.0	−69 08 07
β Beta		3.05	B3	12 46 16.9	−68 06 29
δ Delta		3.62	K2	13 02 16.3	−71 32 56
λ Lambda		3.64	A5	11 45 36.4	−66 43 43
γ Gamma		3.87	B5	12 32 28.1	−72 07 58

Variable stars

	RA h m	Dec. ° '	Range (mags.)	Type	Period (d)	Spectrum
S	12 12.8	−70 09	5.9–6.4	Cepheid	9.66	F
BO	12 34.9	−67 45	6.0–6.7	Irreg.		M
R	12 42.1	−69 24	5.9–6.7	Cepheid	7.48	F
T	13 21.2	−74 27	7.0–9.0	Semi-reg.	93	N

Double stars

	RA h m	Dec. ° '	PA °	Sep. sec	Mags.
β	12 46.3	−68 06	014	1.4	3.7, 4.0
θ	13 08.1	−65 18	187	5.3	5.7, 7.3

Open cluster

M/C	NGC	RA h m	Dec. ° '	Diameter min.	Mag.	No of stars
	4463	12 30.0	−64 48	5	7.2	30

Globular clusters

M/C	NGC	RA h m	Dec. ° '	Diameter min.	Mag.
C108	4372	12 25.8	−72 40	18.6	7.8
C105	4833	12 59.6	−70 53	13.5	7.3

CHAMÆLEON

No star above magnitude 4

Variable stars

	RA h m	Dec. ° '	Range (mags.)	Type	Period (d)	Spectrum
R	08 21.8	−76 21	7.5–14.2	Mira	334.6	M
RS	08 43.2	−79 04	6.0–6.7	Algol + δ Scuti	1.67	A–F

145

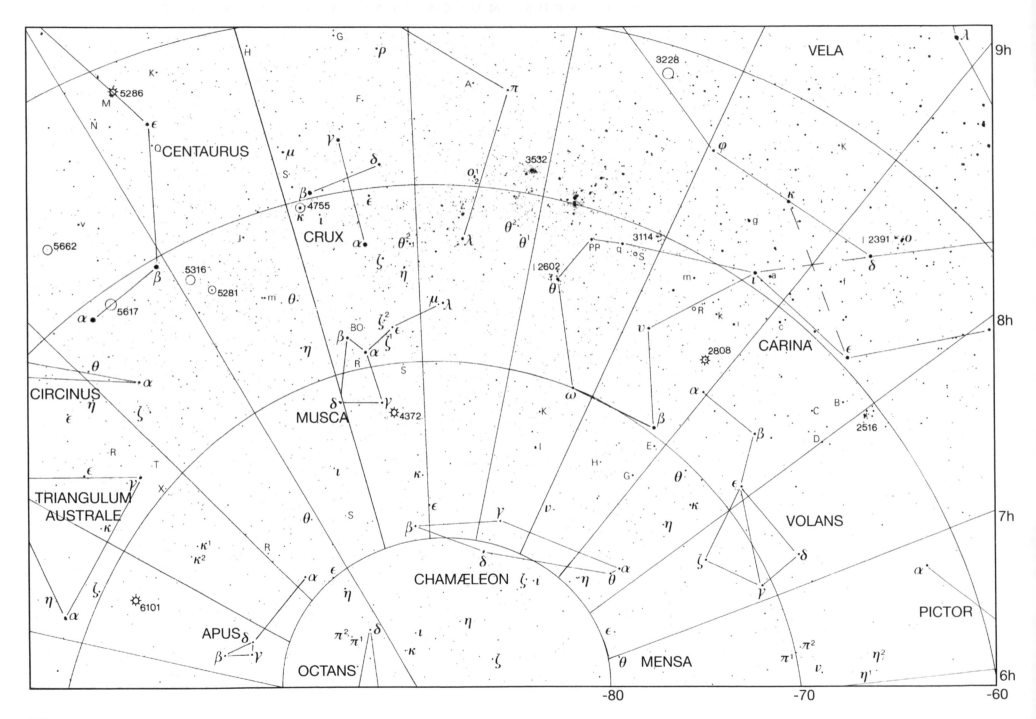

Map 33

URSA MINOR, DRACO

This map contains the North Celestial Pole and the whole of Ursa Minor and Draco, with parts of Cepheus, Camelopardalis, Ursa Major, Hercules and Boötes.

Ursa Minor conjures up the impression of a dim and distorted Ursa Major. Polaris is within one degree of the polar point, and is being moved even closer; on 24 March 2100 its declination will be +89° 32' 51". Subsequently, the pole will move away, and in 12,000 years' time the nearest really bright star to the polar point will be Vega.

Polaris is a luminous star, 6,000 times as powerful as the Sun and 680 light-years away. It is of type F8. It is a pulsating variable with a very small range – magnitude 1.92 to 2.07 according to the Cambridge catalogue – but there is evidence that the amplitude is decreasing, and it has even been suggested that in the near future the pulsations may stop altogether, which would represent a real change in the star's evolution. Whether or not this will happen remains to be seen. Polaris has a faint companion which shares its motion through space, so that the two are presumably associated, but they are a very long way apart. The companion is a useful test for a telescope of around 5-centimetre (2-inch) aperture.

The other bright star in Ursa Minor is Beta or Kocab (2.1, K4); it and its neighbour Gamma (3.0, A3) are often nicknamed the Guardians of the Pole. Beta is strongly orange, as seen from the photograph; it is 95 light-years away and 110 times as luminous as the Sun. Gamma makes up a neat little pair with 11 Ursæ Minoris (5.1, K4), which, like Beta, is orange; the colour comes out well in the photograph even though the star is faint. The other stars in the main pattern of the Little Bear are Delta (4.4, A1); Epsilon (4.2, G5); Zeta (4.3, A3) and

Eta (4.9, F0). Theta (5.3, K4), near Zeta, is yet another orange star.

Several variables in Ursa Minor appear on the photograph. V, above and right of Beta in the photograph, is a semi-regular of type M; at the time the picture was taken the magnitude was not far below 7, and the colour is detectable. Other Mira stars are RR and U, more or less between the Guardians and Draco; RR is easy to locate by its colour, but U was well below maximum at the time of the photograph, and is only just detectable.

Draco is a long, winding constellation. The most prominent part is the Dragon's head near Vega (which is just off the map to the left, below centre). The head is made up of Gamma (2.2, K5); Beta (2.8, G2); Xi (3.7, K2) and Nu, which is a very wide double with components which are virtually equal at magnitude 4.9, and are easily separable with binoculars. They make up a physically-associated system, but the real distance between them is at least 290,000 kilometres (180,000 miles); both components are white, and of type A. Both Gamma and Xi are orange, but the colour of Gamma is the more obvious simply because the star is more than a magnitude the brighter of the two. Just above Xi on the photograph is the very red Mira variable T Draconis; it has an N-type spectrum, but was well away from maximum at the time of the photograph, so that it does not show up well. At its best it can attain almost the seventh magnitude.

Another Mira star, R Draconis, lies to the right of Zeta (3.2, B6); it too is very red, and is well shown. Identification is made easier by the nearby pair of stars,

of which the brighter is 15 Draconis (5.0, B9). Also of type N is RY Draconis, not far from the little trio of which Kappa (3.9, B7) is the brightest member. RY never fades below the 8th magnitude. Yet another telescopic variable is UX Draconis, not far from Epsilon (3.8, G8); it lies close to 59 Draconis (5.1, A8). Epsilon was once included in lists of suspected variables, and in a catalogue drawn up in the late nineteenth century by G.F. Chambers it is listed as fluctuating between magnitudes 3.75 and 4.75. This has never been confirmed, and any real change is most unlikely, but it may be compared with the orange Tau (4.4, K3).

Epsilon has a companion of magnitude 7.4; other easy telescopic doubles are Eta (2.7, G8) and Psi (4.6, F5). Mu (4.9, F5) is a binary with equal components, but this is a difficult object, since the separation is below 2 seconds of arc.

Thuban, or Alpha Draconis (3.6, A0) used to be the north pole star at the time when the Pyramids were built; it is itself undistinguished – 230 light-years away, and 130 times as luminous as the Sun. Close beside it is 10 Draconis (4.7, M3) which is extremely red; the two stars are in the same binocular field, and provide striking colour contrast.

The planetary nebula NGC 6543 (C6) lies near Omega (4.8, F5); it can just be traced on the photograph. To its left on the picture are 42 Draconis (5.0, K1) and 36 (5.0, F3). C6 is one of the brighter planetaries, with an integrated magnitude of above 9 and a central star only a magnitude fainter. Photographs show that its structure is decidedly complex. It has the distinction of being the first nebular object to be examined spectroscopically, by Huggins on 29 August 1864; as soon as Huggins saw the spectrum he realized that he was dealing with a gaseous object rather than a mass of closely-packed stars. The real diameter may be as much as a third of a light-year. In a small telescope it looks like a faint star somewhat out of focus. Incidentally, it lies very near the north pole of the ecliptic.

A small part of Ursa Major is shown here, including the Pinwheel Galaxy, M101, which is just off the photograph illustrating Map 24. It was yet another Méchain discovery, in 1781, and is a loose spiral. It is less than 12 million light-years away, so that it is not so very far beyond the Local Group. It forms almost an equilateral triangle with Eta and Zeta Ursæ Majoris, which are beyond the limit of this photograph but are shown with Map 24.

At the bottom left of the photograph is M92, a globular cluster in Hercules. It is on the fringe of naked-eye visibility, and is not much inferior to M13, the more famous of the two Hercules globulars.

URSA MINOR

Brightest stars

Star		Proper name	Mag.	Spectrum	RA h m s	Dec. ° ' "
α	Alpha	Polaris	1.99v	F8	02 31 50.4	+89 15 51
β	Beta	Kocab	2.08	K4	14 50 42.2	+74 09 19
γ	Gamma	Pherkad Major	3.05	A3	15 20 43.6	+71 50 02

Variable stars

	RA h m	Dec. ° '	Range (mags.)	Type	Period (d)	Spectrum
V	13 38.7	+74 19	7.4–8.8	Semi-reg.	72	M
U	14 17.3	+66 48	7.4–12.7	Mira	326.5	M
RR	14 57.6	+65 56	6.0–6.5	Semi-reg.?	40?	M

Double star

	RA h m	Dec. ° '	PA °	Sep. sec	Mags.
α	02 31.8	+89 16	218	18.4	2.0, 9.0

DRACO

Brightest stars

Star		Proper name	Mag.	Spectrum	RA h m s	Dec. ° ' "
γ	Gamma	Eltamin	2.23	K5	17 56 36.2	+51 29 20
η	Eta	Aldhibain	2.74	G8	16 23 59.3	+61 30 50
β	Beta	Alwaid	2.79	G2	17 30 25.8	+52 18 05
δ	Delta	Taïs	3.07	G9	19 12 33.1	+67 39 41
ζ	Zeta	Aldhibah	3.17	B6	17 08 47.0	+65 42 53
ι	Iota	Edasich	3.29	K2	15 24 55.6	+58 57 58
χ	Chi		3.57	F7	18 21 03.0	+72 43 58
α	Alpha	Thuban	3.65	A0	14 04 23.2	+64 22 33
ξ	Xi	Juza	3.75	K2	17 53 31.5	+56 52 21
ε	Epsilon	Tyl	3.83	G8	19 48 10.2	+70 16 04
λ	Lambda	Giansar	3.84	M0	11 31 24.2	+69 19 52
κ	Kappa		3.87	B7	12 33 28.9	+69 47 17

Variable stars

	RA h m	Dec. ° '	Range (mags.)	Type	Period (d)	Spectrum
RY	12 56.4	+66 00	5.6–8.0	Semi-reg.	173	N
R	16 32.7	+66 45	6.7–13.0	Mira	245.5	M
T	17 56.4	+58 13	7.2–13.5	Mira	421.2	N
UW	17 57.5	+54 40	7.0–8.0	Irreg.		K
UX	19 21.6	+76 34	5.9–7.1	Semi-reg.	168	N

Double stars

	RA h m	Dec. ° '	PA °	Sep. sec	Mags.	
η	16 24.0	+61 31	142	5.2	2.7, 8.7	
μ	17 05.3	+54 28	020	1.9	5.7, 5.7	Binary, 482 y
ν	17 32.2	+55 11	312	61.9	4.9, 4.9	
ψ	17 41.9	+72 09	015	30.3	4.9, 6.1	
ε	19 48.2	+70 16	015	3.1	3.8, 7.4	

Planetary nebula

M/C	NGC	RA h m	Dec. ° '	Dimensions sec.	Mag.	Mag. of central star
C6	6543	17 58.6	+66 38	18 × 350	8.8	9.5

Galaxies

M/C	NGC	RA h m	Dec. ° '	Mag.	Dimensions min.	Type
	4125	12 08.1	+65 11	9.8	5.1 × 3.2	E5p
C3	4236	12 16.7	+69 28	9.7	18.6 × 6.9	Sb
	5866	15 06.5	+55 46	10.0	5.2 × 2.3	E6p

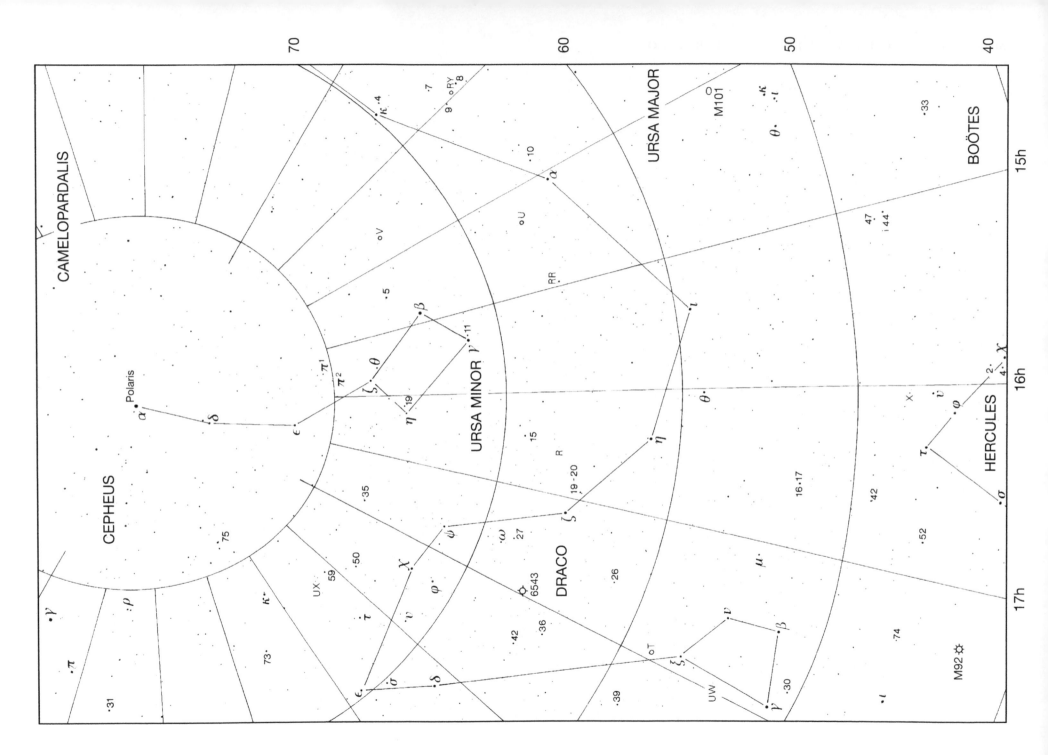

Map 34

BOÖTES, CORONA BOREALIS

This map contains the whole of Boötes and Corona Borealis, with parts of Draco, Ursa Major, Canes Venatici, Coma, Virgo, Serpens and Hercules. Note the two famous galaxies M51 (the Whirlpool) in Canes Venatici and M101 (the Pinwheel) in Ursa Major; they lie on opposite sides of the "tail" of the Great Bear. Eta Ursæ Majoris (Alkaid) is shown on the map, near the upper right-hand edge, but Zeta (Mizar) is just out of view; it is shown, with the rest of Ursa Major, in Map 24.

Arcturus is of magnitude –0.04. It is surpassed only by Sirius, Canopus and Alpha Centauri, all of which are south of the celestial equator, so that Arcturus is the brightest star in the northern hemisphere of the sky. It is of a lovely light orange colour, and it is not easy to see why seamen of ancient times regarded it as unlucky; Pliny even calls it "horridum sidus".

It ranks as a giant star, though it cannot be compared with Betelgeux or Antares. It is well over 100 times as luminous as the Sun, and its diameter is of the order of 19 million kilometres (12 million miles); its distance is 36 light-years. It has been found that the amount of heat it sends to us is about the same as that received from a single candle 8 kilometres (5 miles) away.

In April 1860, J Baxendell, a leading astronomical observer of the time, noted a star of above the tenth magnitude in the same telescopic field as Arcturus. It gradually faded away, and by the end of the month had disappeared altogether. Though it has been given an official variable star designation, T Boötis, it has never been seen again. It may have been a recurrent nova, in which case it may one day reappear; certainly there is no harm in checking. Modern photographs show no star as bright as magnitude 17 in the appropriate position.

The other main stars of the Boötes pattern are Epsilon (2.4, K0); Delta (3.5, G8); Beta (3.5, G8); Gamma (3.0, A7); Rho (3.6, K3); Eta (2.7, G0) and Zeta (3.8, A2). On the photograph both Epsilon and Rho show up as slightly orange, but the colours are not at all pronounced. Epsilon is an easy double; the companion, of magnitude 4.9, is often said to be greenish, but this is probably due to contrast with the primary. The two definitely form a physical pair, but they are a long way apart, and the orbital period must amount to thousands of years. Zeta is a binary with equal components and a period of 123 years, but as the separation is only one second of arc it is by no means an easy object. Mu (4.3, F0) has a distant companion with the same motion through space; Xi (4.6, G8) is a well-known binary, separable with a small telescope. It is easy to locate on the map, well above Zeta.

44 Boötis is another binary – actually an eclipsing system, though the magnitude range is small. It lies well above Beta on the map, close beside 47 (5.6, A0); running below it is a neat little curve of faint stars, well seen with binoculars. The primary of 44 Boötis is very similar to the Sun in size, type and luminosity. The period is 225 years, and the apparent orbit is a very elongated ellipse, so that the separation ranges between almost 5 seconds of arc (as in 1880) to only 0.4 of a second of arc. The apparent separation was least in 1969, and is now increasing again. Kappa, Iota and Pi are easy doubles.

Two variables lie close to Epsilon. One, W Boötis, comes out well in the photograph; it is a red semi-regular, with a range of from magnitude 4.7 to 5.4. The other, R Boötis, is a Mira star. It can reach almost the sixth magnitude, but was well away from maximum at the time of the photograph, and is only just detectable; at minimum it falls below magnitude 13. Another red variable, V Boötis (near Gamma) was also faint when the photograph was taken, but when it is at its best, around magnitude 7, its colour is very pronounced.

Beta Boötis lies near the radiant of the Quadrantid meteor shower, seen every year in early January. The constellation Quadrans was added to the sky by JE Bode, in his maps of 1775, but has long since been rejected. In view of the fact that its name is still attached to the January meteors, this is perhaps rather a pity!

A prominent Y-pattern is formed by Arcturus, Epsilon Boötis, Gamma Boötis and Alpha Coronæ, the leader of the little constellation of the Northern Crown. Alpha Coronæ, often referred to by its proper name of Alphekka, is of magnitude 2.2 and type A0 (it is actually variable, but with a range of no more than a tenth of a magnitude). Corona itself is very distinctive, though apart from Alphekka its leading stars are not bright; they range from magnitude 3.7 (Beta) to 5.0 (Iota).

The "bowl" of the Crown contains R Coronæ, one of the most interesting variables in the sky. Ordinarily it is of around magnitude 6, on the fringe of naked-eye visibility; as in the photograph; compare it with the 6.6 magnitude star on the opposite side of the bowl near Theta Coronæ (4.1, B5). At unpredictable intervals R Coronæ fades, falling to below magnitude 14 and passing beyond the range of small telescopes. What apparently happens is the star accumulates clouds of soot in its atmosphere, and these mask the light coming from below; minimum ends only when the clouds are dispersed. R Coronæ variables are deficient in hydrogen, but rich in carbon; only two attain naked-eye visibility – R Coronæ itself, and RY Sagittarii (Map 43).

R Coronæ is a favourite binocular object. If you look at the bowl and see only one reasonably prominent star, you may be sure that R Coronæ is passing through minimum. It may remain faint for weeks or even months before returning to normal.

Just outside the bowl is the Blaze Star, T Coronæ, which is normally of around the tenth magnitude, and therefore too faint to be shown in this photograph. However, in 1866 and again in 1946 it rose briefly to naked-eye visibility, and may well do so again in the future. It is the classic example of a recurrent nova.

The Mira variable S Coronæ is shown here, between Eta (5.0, G0) and the little pair of stars closely to the left of Delta Boötis. When this photograph was taken, the star was well below its maximum magnitude of 5.8.

There are various other doubles in Corona. Eta has almost equal components, but the separation is only about 1 second of arc. Above the bowl in the photograph, Rho has a ninth-magnitude companion; Sigma is an easy binary. Another good double is Zeta, where the separation is over 6 seconds of arc. Above and to the right of Zeta is a neat little curved line of stars, and an easily-recognizable pair, Nu¹ and Nu², both of which are of the fifth magnitude.

BOÖTES

Brightest stars

Star		Proper name	Mag.	Spectrum	RA h m s	Dec. ° ' "
α	Alpha	Arcturus	-0.04	K2	14 15 39.6	+19 10 57
ε	Epsilon	Izar	2.37	K0	14 44 59.1	+27 04 27
η	Eta		2.68	G0	13 54 41.0	+18 23 51
γ	Gamma	Seginus	3.03	A7	14 32 04.6	+38 13 30
δ	Delta		3.47	G8	15 15 30.1	+33 18 53
β	Beta	Nekkar	3.50	G8	15 01 56.6	+40 23 26
ρ	Rho		3.58	K3	14 31 49.7	+30 22 17
ζ	Zeta		3.78	A2	14 41 08.8	+13 43 42

Variable stars

	RA h m	Dec. ° '	Range (mags.)	Type	Period (d)	Spectrum
V	14 29.8	+38 52	7.0–12.0	Semi-reg.	258	M
R	14 37.2	+26 44	6.2–13.1	Mira	223.4	M
W	14 43.4	+26 32	4.7–5.4	Semi-reg.	450	M
i (44)	15 03.8	+47 39	6.5–7.1	W UMa	0.27	G+G

Double stars

	RA h m	Dec. ° '	PA °	Sep. sec	Mags.	
κ	14 13.5	+51 47	236	13.4	4.6, 6.6	
ι	14 16.2	+51 22	033	38.5	4.9, 7.5	
π	14 40.7	+16 25	108	5.6	4.9, 5.8	
ζ	14 41.1	+13 44	AB 303	1.0	4.5, 4.6	Binary, 123.3 y
			AC 259	99.3	10.9	
i (44)	15 03.8	+47 39	040	1.0	5.3v, 6.2	Binary, 225 y
μ	15 24.5	+37 23	171	108.3	4.3, 7.0	
ε	14 45.0	+27 04	339	2.8	2.5, 4.9	
ξ	14 51.4	+19 06	095	282	4.7, 6.8	Binary, 150 y

Galaxy

M/C	NGC	RA h m	Dec. ° '	Mag.	Dimensions min.	Type
C45	5248	13 37.5	+08 53	10.2	6.5 × 4.9	Sc

CORONA BOREALIS

Brightest stars

Star		Proper name	Mag.	Spectrum	RA h m s	Dec. ° ' "
α	Alpha	Alphekka	2.23v	A0	15 34 41.2	+26 42 53
β	Beta	Nusakan	3.68	F0p	15 27 49.7	+29 06 21
γ	Gamma		3.84v	A0	15 42 44.5	+26 17 44

Variable stars

	RA h m	Dec. ° '	Range (mags.)	Type	Period (d)	Spectrum
U	15 18.2	+31 39	7.7–8.8	Algol	3.45	B+F
S	15 21.4	+31 22	5.8–14.1	Mira	360.3	M
R	15 48.6	+28 09	5.7–15	R CrB		F8p
V	15 49.5	+39 34	6.9–12.6	Mira	357.6	N
T	15 59.5	+25 55	2.0–10.8	Recurrent nova		M+Q

(Outbursts 1866, 1946)

Double stars

	RA h m	Dec. ° '	PA °	Sep. sec	Mags.	
o	15 20.1	+29 37	337	147.3	5.5, 9.4	
η	15 23.2	+30 17	AB 030	1.0	5.6, 5.9	Binary, 41.6 y
			AC 012	57.7	12.5	
			AB+D 047	215.0	10.0	
ζ	15 39.4	+36 38	305	6.3	5.1, 6.0	
ρ	16 01.0	+33 18	071	89.8	5.5, 8.7	
σ	16 14.7	+33 52	234	7.0	5.6, 6.6	Binary, 1000 y

URSA MAJOR

Galaxy

M/C	NGC	RA h m	Dec. ° '	Mag.	Dimensions min.	Type
M101 (Pinwheel)	5457	14 03.2	+54 21	7.7	26.9 × 26.3	Sc

The whole of Ursa Major is shown on Map 24, and objects are listed there.

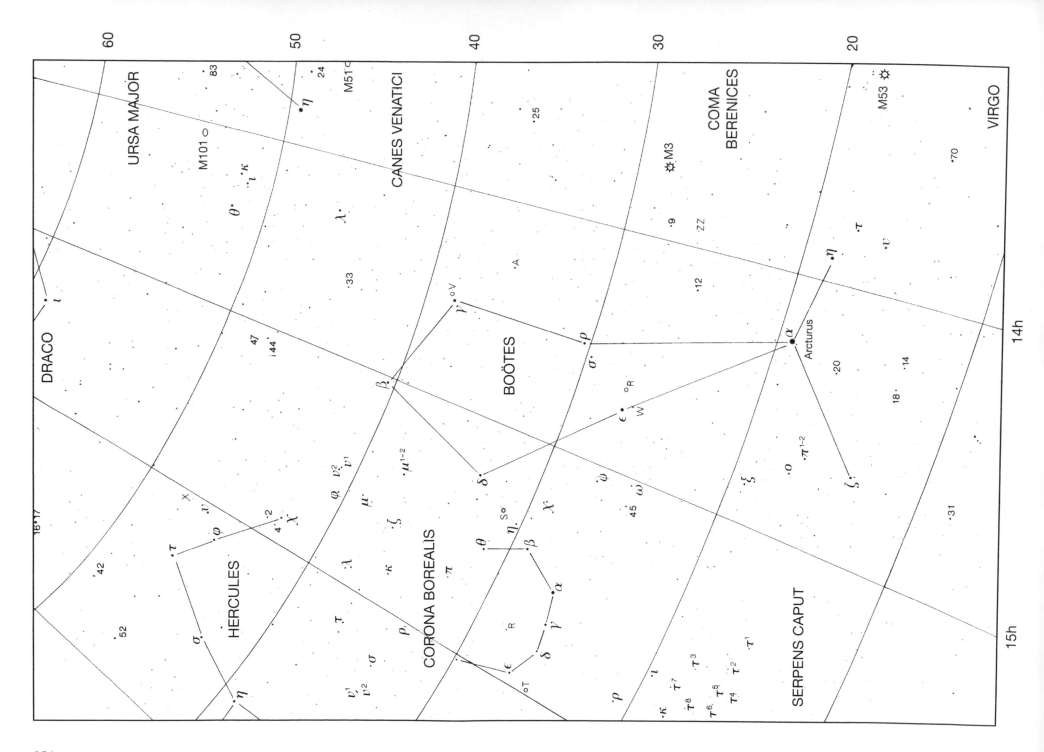

Map 35

HERCULES

Almost the whole of this map is occupied by Hercules, named in honour of the mythological hero; he covers a large area, but is not so imposing as might be expected! The map also includes the head of Draco, with the orange Gamma Draconis prominent near the top of the photograph; a few stars of Lyra, including the brilliant Vega; the whole of Corona Borealis, and small parts of Serpens and Ophiuchus.

The most interesting star in Hercules is Alpha, often known by its proper name of Rasalgethi. It is rather divorced from the rest of the constellation, and is seen to the lower left in the photograph, to the right of the rather brighter Alpha Ophiuchi or Rasalhague (2.1, A5). Alpha Herculis is a red supergiant. On this image it appears at least as bright as the other two leaders of the constellation, Beta (2.8, G8) and Zeta (2.8, G0). It is a semi-regular variable; the official range is between magnitudes 3 and 4, but for most of the time it remains between 3.1 and 3.5. The period is said to be of the order of 100 days, but this is at best very ill-defined. The most useful naked-eye comparison star is Kappa Ophiuchi (3.2, K2), which is just visible at the very bottom of the photograph; it makes a pair with Iota Ophiuchi (4.4, B8).

Alpha Herculis has an M-type spectrum; it is a huge star, with a relatively low surface temperature. According to the Cambridge catalogue it is 220 light-years away and, on average, 700 times as luminous as the Sun, in which case it cannot be compared with a true colossus such as Betelgeux. It must be added, however, that some other catalogues make it more remote and more powerful than the Cambridge values.

It has a companion of magnitude 5.4; the separation is almost 5 seconds of arc. The companion has been variously described as blue, green or emerald; no doubt the proximity of the red primary has a great deal to do with this. In any case, Alpha Herculis is a beautiful sight when seen through an adequate telescope. The two components are genuinely associated; it has been estimated that the orbital period is around 3,600 years. Finally, it is worth noting that if our eyes were sensitive to infra-red radiation, Alpha Herculis would be one of the most brilliant objects in the entire sky.

Zeta Herculis is a fine binary; its double nature was discovered by William Herschel as long ago as 1782. The primary is a G-type sub-giant, 5 times as luminous as the Sun, while the K-type dwarf companion has about half the Sun's luminosity. The distance from us is 31 light-years, and the orbital period is only 34.5 years, so that the separation and position angle alter quite rapidly; the present separation is 1.6 seconds of arc.

Rho (4.2, A0), close to the brighter orange Pi (3.2, K3) is an easy telescopic double; so too is Kappa (5.0, K2), well below and to the right of Gamma (3.7, A9), Gamma is itself a wide double, though the companion is inconveniently faint.

Delta Herculis (3.1, A3), well above and rather to the left of Alpha, has a companion of magnitude 8. This is an excellent example of an optical pair; there is no connection between the two components, and in fact they are moving through space in completely different directions. At present, the separation is nearly 9 seconds of arc, but the two are moving away from each other, so that the separation will continue to increase indefinitely. The secondary is of type G4, and there have been discordant estimates of its colour, though no doubt most observers will call it white.

68 (u) Herculis, which forms a triangle with Epsilon (3.9, A0) and Pi, is an interesting eclipsing binary; it is a Beta Lyræ type system, made up of two B-type giants orbiting each other in a period of just over 2 days. As with all systems of this kind, the two are very close together; centre to centre the separation cannot be much more than 9.7 million kilometres (6 million miles). The magnitude range is from 4.6 to 5.3, so that this is a suitable object for binocular observers.

Several red variables in Hercules are identifiable on the photograph. U, near Gamma, is just detectable, though it was well below its maximum magnitude of 6.5; X, to the right of Tau (3.9, B5) was close to its maximum of magnitude 7.5, so that its colour comes out well; T, close to the triangle made up of Xi (3.7, K0), Nu (4.4, F2) and Omicron (3.8, B9) was almost at minimum, and difficult to see here; RU, between Beta Herculis and Epsilon Coronæ, can be seen; so can S, to the right of Alpha in the image, beyond a characteristic curved line of faint stars. A more prominent variable is g (30), below and to the right of Sigma (4.2, B9). The range is from magnitude 4 to 6, and at the time of the photograph the star was almost at maximum; indeed, it appears little inferior to Sigma. It is semi-regular, with a mean period of about 70 days.

Of course, the most important objects in Hercules are the two globular clusters, M13 and M92. M13 lies between Zeta and Eta, and is much the finer; it is distinctly visible with the naked eye under good conditions, and is shown on the photograph, though it looks very like a star. It was discovered by Halley in 1714; it is around 22,000 light-years away, with a real diameter of 160 light-years. It is very well seen with binoculars, and the outer parts are not hard to resolve into stars – though, strangely, it is very poor in short-period variables.

M92, discovered by JE Bode in 1777, is telescopically not greatly inferior to M13, though it is fainter and further away (about 37,000 light-years). It can just be seen with the naked eye on a very dark, clear night. It is rather more condensed than M13, but a modest telescope will resolve its outer portions without much difficulty.

HERCULES

Brightest stars

Star		Proper name	Mag.	Spectrum	RA h m s	Dec. ° ' "
β	Beta	Kornephoros	2.77	G8	16 30 13.1	+21 29 22
ζ	Zeta	Rutilicus	2.81	G0	16 41 17.1	+31 16 10
α	Alpha	Rasalgethi	3.0 max.	M5	17 14 38.8	+14 23 25
δ	Delta	Sarin	3.14	A3	17 15 01.8	+24 50 21
π	Pi		3.16	K3	17 15 02.6	+36 48 33
μ	Mu		3.42	G5	17 46 27.3	+27 43 15
η	Eta		3.53	G8	16 42 53.7	+38 55 20
ξ	Xi		3.70	K0	17 57 45.7	+29 14 52
γ	Gamma		3.75	A9	16 21 55.1	+19 09 11
ι	Iota		3.80	B3	17 39 27.7	+46 00 23
o	Omicron		3.83	B9	18 07 32.4	+28 45 45
	109		3.84	K2	18 23 41.7	+21 46 11
θ	Theta		3.86	K1	17 56 15.1	+37 15 02
τ	Tau		3.89	B5	16 19 44.3	+46 18 48
ε	Epsilon		3.92	A0	17 00 17.2	+30 55 35

Variable stars

	RA h m	Dec. ° '	Range (mags.)	Type	Period (d)	Spectrum
X	16 02.7	+47 14	7.5–8.6	Semi-reg.	95	M
R	16 06.2	+18 22	7.8–15	Mira	318.4	M
RU	16 10.2	+25 04	6.8–14.3	Mira	485.5	M
U	16 25.8	+18 54	6.5–13.4	Mira	406	M
g (30)	16 28.6	+41 53	5.7–7.2	Semi-reg.	70	M
S	16 51.9	+14 56	6.4–13.8	Mira	307.4	M
α	17 14.6	+14 23	3.0–4.0	Semi-reg.	±100?	M
u (68)	17 17.3	+35 06	4.6–5.3	Beta Lyræ	2.05	B+B
RS	17 21.7	+22 55	7.0–13	Mira	219.6	M
Z	17 58.1	+15 08	7.3–8.1	Algol	3.99	F+K
T	18 09.1	+31 01	6.8–13.9	Mira	165	M
AC	18 30.3	+21 52	7.4–9.7	RV Tauri	75.5	F+K

Double stars

	RA h m	Dec. ° '	PA °	Sep. sec	Mags.	
κ	16 08.1	+17 03	012	28.4	5.3, 6.5	
γ	16 21.9	+19 09	233	41.6	3.8, 9.8	
37	16 40.6	+04 13	230	69.8	5.8, 7.0	
ζ	16 41.3	+31 36	089	1.6	2.9, 5.5	Binary, 34.5 y
α	17 14.6	+14 23	107	4.7	var., 5.4	Binary, 3600 y
δ	17 15.0	+24 50	236	8.9	3.7, 8.2	Optical pair
ρ	17 23.7	+37 09	316	4.1	4.6, 5.6	

Globular clusters

M/C	NGC	RA h m	Dec. ° '	Diameter min.	Mag.
M92	6341	17 17.1	+43 08	11.2	6.5
M13	6205	16 41.7	+36 28	16.6	5.9

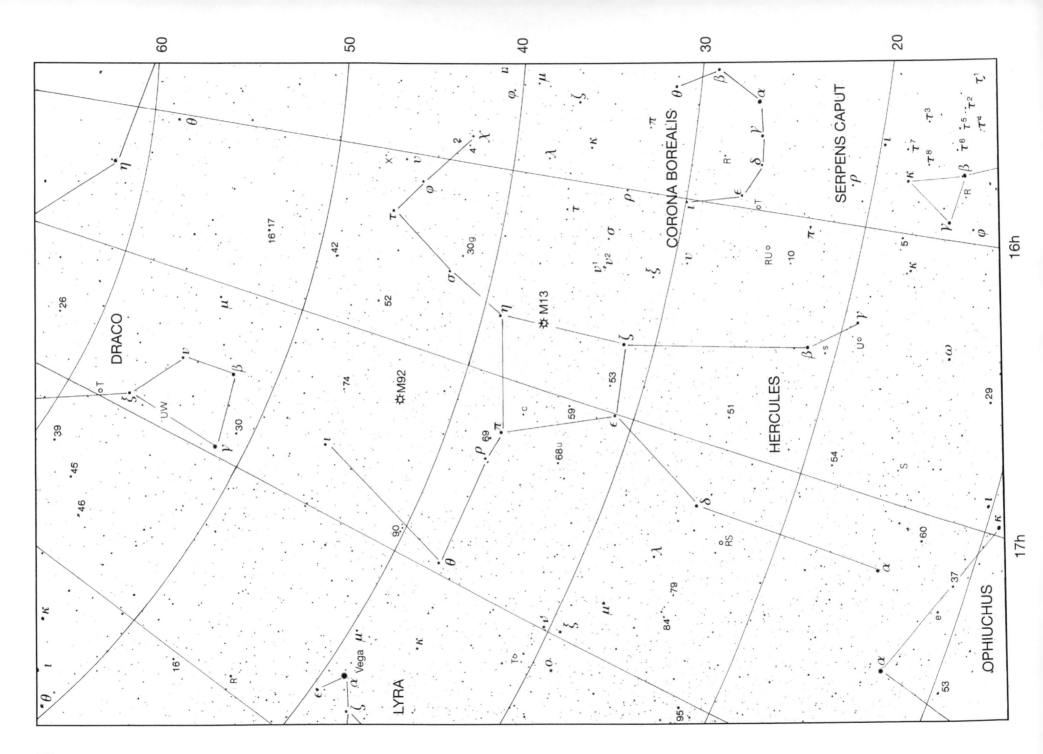

Map 36
OPHIUCHUS, SERPENS

Ophiuchus dominates this map; like Hercules, it covers a wide area, but has no well-defined pattern, though it does include one bright star, Alpha or Rasalhague (2.1, A5). Alpha Ophiuchi is shown to the left-hand side of the photograph; above it, to its right, is the red variable Alpha Herculis. Owing to its variability Alpha Herculis appears to be slightly fainter than Beta Herculis, whereas on the photograph with Map 35 they appear more or less equal.

There are some rich star-fields in this region, and the Milky Way crosses the map at the bottom left. Most of Serpens is also shown – both the head (Caput) and the body (Cauda), which are separated by Ophiuchus. It seems that the serpent and the serpent-bearer have been locked in deadly combat, and that the luckless reptile has been having the worst of the encounter! A few stars of Corona Borealis appear at the extreme upper left of the photograph.

Delta Ophiuchi (2.7, M1), near the right-hand edge of the photograph, is a beautiful red star; it contrasts sharply with its neighbour Epsilon (3.2, G8). At its distance of only 140 light-years, Delta is one of the nearest of the red giants, and attempts are being made to detect surface features on it, as has already been possible with Betelgeux in Orion.

One much fainter star of unusual interest is shown on the photograph: Barnard's Star, discovered in 1916 by the American astronomer Edward Emerson Barnard (its official designation is Munich 15040). It is 6 light-years away, so that it is the nearest of all stars beyond the Sun apart from those of the Alpha Centauri group. It is feeble even by cosmical standards, with luminosity no more

than 0.00045 that of the Sun. Its annual proper motion is 10.3 seconds of arc, so that in around 180 years it crawls across the sky by a distance equal to the apparent diameter of the full moon; its magnitude is 9.5. Some years ago P. van de Kamp claimed that slight irregularities in its movements indicated the existence of one or even two orbiting planets; it was later found that van de Kamp's measurements were faulty, and we have no proof that Barnard's Star is the centre of a planetary system, though the possibility certainly cannot be ruled out.

To identify it, first locate Beta Ophiuchi (2.8, K2) and its neighbour Gamma (3.8, A0). Below and to the left of Gamma are two fainter stars, 67 Ophiuchi (4.0, B5) and 70 (4.0, K0); 70 Ophiuchi is orange. Above and to the left of 67 is 66 Ophiuchi (4.6, B2), and close to this, below a tiny triangle of dim stars, is the faint speck marking Barnard's Star. On the map its position is indicated by bars.

70 Ophiuchi is itself a famous binary, discovered by William Herschel in 1779. The components are of magnitudes 4 and 6, and since the present separation is 1.5 seconds of arc this is not a difficult pair; the orbital period is 88 years, and at times, as in the early 1930s, the apparent separation can approach 7 seconds of arc. The primary is orange, of type K; the secondary is also of type K, but estimates of its colour are rather discordant. The total luminosity of the system is about half that of the Sun.

(As an aside: the little asterism of 66, 67, 68 and 70 Ophiuchi was once separated off as an independent constellation: Taurus Poniatowski (Poniatowski's Bull), proposed in 1777 by an astronomer named Poczobut. It

survived briefly on some maps, but was soon abandoned.)

Several of the main stars in Ophiuchus are doubles. Among these is Eta, where the components are almost equal (magnitudes 3.0 and 3.5). The orbital period is 84.3 years. The separation is only 0.5 of a second of arc, so that this is far from being an easy pair; the average real separation between the components is about the same as that between the Sun and Uranus, but the orbit is highly eccentric; Periastron is next due in 2020.

Tau Ophiuchi (4.8, F0) is a fairly easy pair with reasonably equal components; it lies close to Nu Ophiuchi (3.3, K0) the orange star near the bottom of the photograph. (In this region, Ophiuchus and Serpens are confusingly mixed.) Yet another easy binary, Lambda (3.8, A1) lies more or less between Kappa, the comparison star for Alpha Herculis, and the red giant Delta.

Y Ophiuchi, above Nu in the photograph, is a Cepheid variable with a range of from magnitude 5.9 to 6.4. To its left is a triangle of stars, of which the brightest member is Zeta Serpentis (4.6, F3). Just to the right of Y is an interesting star, RS Ophiuchi, which is normally very faint (around magnitude 12) but can sometimes flare up to naked-eye visibility, as it did in 1933, 1958 and 1967. It is an excellent example of a recurrent nova, and actually the brightest of its class with the exception of the "Blaze Star" T Coronæ, which is also on this map very near the top (it is better shown in Map 34). It can just be identified on the photograph; when it will next explode remains to be seen, but it is always worth monitoring.

R Ophiuchi, a typical Mira star, can be made out *Continued on page 164*

OPHIUCHUS

Brightest stars

Star		Proper name	Mag.	Spectrum	RA h m s	Dec. ° ' "
α	Alpha	Rasalhague	2.08	A5	17 34 55.9	+12 53 36
η	Eta	Sabik	2.43	A2	17 10 22.5	−15 43 30
ζ	Zeta	Han	2.56	O9.5	16 37 09.4	−10 34 02
δ	Delta	Yed Prior	2.74	M1	16 14 20.6	−03 41 39
β	Beta	Cheleb	2.77	K2	17 43 28.2	+04 34 02
κ	Kappa		3.20	K2	16 57 40.0	+09 22 30
ε	Epsilon	Yed Post	3.24	G8	16 18 19.1	−04 41 33
θ	Theta		3.27	B2	17 22 00.4	−24 59 58
ν	Nu		3.34	K0	17 59 01.4	−09 46 25
	72		3.73	A4	18 07 20.8	+09 33 50
γ	Gamma		3.75	A0	17 47 53.4	+02 42 26
λ	Lambda	Marfik	3.82	A1	16 30 54.7	+01 59 02
	67		3.97	B5	18 00 38.5	+02 55 53

Variable stars

	RA h m	Dec. ° '	Range (mags.)	Type	Period (d)	Spectrum
V	16 26.7	−12 26	7.3–11.6	Mira	298	N
χ	16 27.0	−18 27	4.2–5.0	Irreg.		B
R	17 07.8	−16 06	7.0–13.8	Mira	302.6	
U	17 16.5	+01 13	5.9–6.6	Algol	1.68	B+B
RS	17 50.2	−06 43	5.3–12.3	Recurrent nova		O+M
(Outbursts 1933, 1958, 1967)						
Y	17 52.6	−06 09	5.9–6.4	Cepheid	17.12	F–G
X	18 38.3	+08 50	5.9–9.2	Mira	334.4	M+K

Double stars

	RA h m	Dec. ° '	PA °	Sep. sec	Mags.	
ρ	16 25.6	−23 27	344	3.1	5.3, 6.0	
λ	16 30.9	+01 59	022	1.5	4.2, 5.2	Binary, 129.9 y
19	16 47.2	+02 04	089	23.4	6.1, 9.4	
36	17 15.3	−26 36	150	4.7	5.1, 5.1	Binary, 549 y
τ	18 03.1	−08 11	AB 280	1.8	5.2, 5.9	Binary, 280 y
			AC 127	100.3	9.3	
70	18 05.5	+02 30	224	1.5	4.2, 6.0	Binary, 88.1 y

Open clusters

M/C	NGC	RA h m	Dec. ° '	Diameter min.	Mag.	No of stars
	IC 4665	17 46.3	+05 43	41	4.2	30
	6633	18 27.7	+06 34	27	4.6	30

Globular cluster

M/C	NGC	RA h m	Dec. ° '	Diameter min.	Mag.
M107	6171	16 32.5	−13 03	10.0	8.1
M12	6218	16 47.2	−01 57	14.5	6.6
M10	6254	16 57.1	−04 06	15.1	6.6
M62	6266	17 01.2	−30 07	14.1	6.6
M19	6273	17 02.6	−26 16	13.5	7.1
M9	6333	17 19.2	−18 31	9.3	7.9
M14	6402	17 37.6	−03 15	11.7	7.6

SERPENS

Brightest stars

Star		Proper name	Mag.	Spectrum	RA h m s	Dec. ° ' "
Caput						
α	Alpha	Unukalhai	2.65	K2	15 44 16.0	+06 25 32
μ	Mu		3.54	A0	15 49 37.1	−03 25 49
β	Beta		3.67	A2	15 46 11.2	+15 25 18
ε	Epsilon		3.71	A2	15 50 48.9	+04 28 40
δ	Delta	Tsin	3.80	F0	15 34 48.0	+10 32 21
γ	Gamma		3.85	F6	15 56 27.1	+15 39 42
Cauda						
η	Eta	Alava	3.26	K0	18 21 18.4	−02 53 56
θ	Theta	Alya	3.4	A5+A5	18 56 14	+04 12 10
		(combined)				
ξ	Xi		3.54	F0	17 37 35.0	−15 23 55

Variable stars

	RA h m	Dec. ° '	Range (mags.)	Type	Period (d)	Spectrum
S	15 21.7	+14 19	7.0–14.1	Mira	368.6	M
τ⁴	15 36.5	+15 06	7.5–8.9	Irreg.		M
R	15 50.7	+15 08	5.1–14.4	Mira	356.4	M
d	18 27.2	+00 12	4.9–5.9	?	?	G+A

Double stars

	RA h m	Dec. ° '	PA °	Sep. sec	Mags.	
δ	15 34.8	+10 32	177	4.4	4.1, 5.2	Binary, 3168 y
ν	17 20.8	−12 51	028	46.3	4.3, 8.3	
d	18 27.2	+00 12	318	3.8	5.3v, 7.6	
θ	18 56.2	+04 12	104	22.3	4.5, 4.5	

Open clusters

M/C	NGC	RA h m	Dec. ° '	Diameter min.	Mag.	No of stars
	6611	18 18.8	−13 47	7	6.0	In M16
	6604	18 18.1	−12 14	2	7.0	30

Globular cluster

M/C	NGC	RA h m	Dec. ° '	Diameter min.	Mag.
S	5904	15 18.6	+02 05	17.4	5.8

Nebula

M/C	NGC	RA h m	Dec. ° '	Dimensions min.	Mag. of illum. star
16	6611	18 18.8	−13 47	35 × 28	6.4
(Eagle Nebula)					

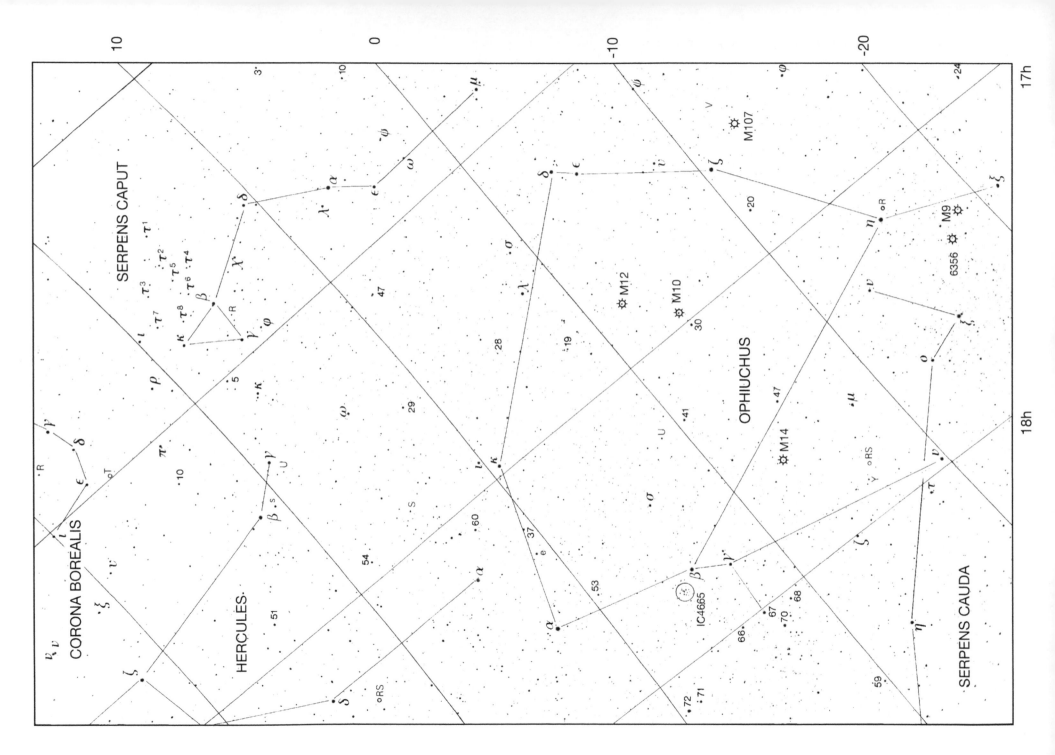

closely to the right of Eta, though at the time of the photograph it was well below its maximum magnitude of 7. V Ophiuchi, to the right of Zeta (2.6, O9.5) is also a Mira variable, but with an N-type spectrum; it can rise to magnitude 7.3, and though it was not at its best when this photograph was taken the colour is still very pronounced. The eclipsing binary U Ophiuchi can also be found, below and to the right of the reddish Sigma (4.3, K3). U Ophiuchi is an Algol star, with a mean magnitude of just below 6. Both components are of type B.

Ophiuchus is very rich in globular clusters, and several can be identified in the photograph, though they are not very bright and show up as dim, slightly blurred stars. M10, above and to the right of 30 Ophiuchi (4.8, K4) is probably the most conspicuous of them; above it on the photograph is M12, which is about the same integrated magnitude (6.6) but is rather less condensed. M107, below and to the right of Zeta, is decidedly fainter. M9, below Eta and very near the bottom of the map, is a small but relatively bright cluster; to its left is another globular, NGC 6356, which is not in Messier's list. There is also M14, in a rather isolated position left of 47 Ophiuchi (4.5, F3). The little line of stars above M14 in the photograph is a help in identification.

Globular-hunting in Ophiuchus can be a diverting pastime. There are plenty of objects from which to choose!

The Serpent's head is much more prominent than its body, and is marked by a triangle near the top of the photograph made up of Beta (3.7, A2); Gamma (3.8, F6) and the very orange Kappa (4.1, M1). The brightest star in Serpens, Alpha (2.6, K2) is clearly orange; it is 85

light-years away and 96 times as luminous as the Sun. Delta (3.8, F0) is an easy double; the separation is 4.4 seconds of arc, and the difference between the components is less than a magnitude (4.1 and 5.2). This is a binary, with a computed period of 3,168 years.

The Mira variable R Serpentis lies directly between Beta and Gamma, which makes it very easy to locate. At its peak it can reach the fifth magnitude, and is a naked-eye object. At the time of the photograph it was nowhere near maximum, but its colour is still very much in evidence. At minimum it sinks below magnitude 14, so that it passes beyond the range of small telescopes. Its period is only nine days less than a year, so that there are times, for several consecutive years, when the maxima are difficult to observe at all.

Tau⁴ Serpentis, at the end of a line of stars extending away from Beta to the upper right, is of type M, and varies between 7.5 and 8.9; its changes are slow and irregular. At the time of the photograph it was almost as bright as it can ever become.

Though almost the whole of Serpens is shown on this map, it so happens that the three most interesting objects in the entire constellation – the globular cluster M5 in Caput, and the double star Theta and the lovely Eagle Nebula 16 in Cauda – are outside its boundaries. Nu Serpentis (4.3, A1) is a wide double, but the companion is below the eighth magnitude.

Map 37

LIBRA

Libra occupies the central part of this map; also shown are parts of Virgo, Centaurus, Serpens, Ophiuchus, Scorpius and Hydra. The brightest stars are Spica (upper right), Antares (bottom left) and the orange Theta Centauri (lower right).

Gamma Hydræ (3.0, G5) is just on the photograph, to the extreme right. Left of it is the red variable R Hydræ; this is one of the brightest of the long-period variables, and was actually the third to be identified, after Mira itself and Chi Cygni. In this photograph the star is some way from maximum, but its red colour is plain; at its best it can attain the 4th magnitude, and never drops below the 10th. The present period is 390 days, but it seems that in the early eighteenth century the period was closer to 500 days, so that we may be watching a real change in the star's evolution. Also in Hydra is the galaxy M83, to the right and slightly below Pi Hydræ (3.3, K2); it is a loose spiral, not so very far beyond the Local Group, and in size very similar to our own system.

Libra is one of the least imposing of the Zodiacal constellations. Its leading stars are Beta (2.6, B8); Alpha² (2.7, A3); the orange Sigma (3.3, M4) and Gamma (3.9, G8). Sigma, with Upsilon (3.6, K5) and Tau (3.7, B4) was once included in Scorpius, as the Scorpion's Claws; on some older maps it appears as Gamma Scorpii. Beta is said to be the only single star with a distinct greenish hue, but most observers will call it white. It is 121 light-years away, and 105 times as luminous as the Sun.

Alpha² makes up a very wide pair with its neighbour Alpha¹; the pair can be well seen with binoculars, since they are over 230 seconds of arc apart. They share a common motion through space, and are presumably

MAP 37 : LIBRA

associated, but the real distance between them cannot be far short of 5,000 astronomical units. Mu (5.4, A4) is an easy double.

Delta Libræ, above and to the right of Beta on the map, is an Algol-type eclipsing binary; it has a range of no more than one magnitude, and a period of 2.3 days. A suitable naked-eye comparison star is Epsilon Libræ (4.9, F5) on the far side of Beta.

48 Libræ (4.9, B3) is a shell star, encircled by an outer ring of gases which occasionally shows violent expansion, causing interesting spectral changes though very little change in light output.

The Scorpion's head is shown here, with the globular cluster M4, close to Antares; Beta is a fine double, while Omega[1] and Omega[2] make up a wide optical pair. Scorpius is better shown with Map 38.

LIBRA

Brightest stars

Star	Proper name	Mag.	Spectrum	RA h m s	Dec. ° ' "
β Beta	Zubenel-chemale	2.61	B8	15 17 00.3	−09 22 58
α² Alpha²	Zubenel-genubi	2.75	A3	14 50 52.6	−16 02 30
σ Sigma	Zubenalgubi	3.29	M4	15 04 04.1	−25 16 55
υ Upsilon		3.58	K5	15 37 01.4	−28 08 06
τ Tau		3.66	B4	15 38 39.3	−29 46 40
γ Gamma	Zubenel-hakrabi	3.91	G8	15 35 31.5	−14 47 23

Variable stars

	RA h m	Dec. ° '	Range (mags.)	Type	Period (d)	Spectrum
δ	15 01.1	−08 31	4.9–5.9	Algol	2.33	B
RS	15 24.3	−22 55	7.0–13.0	Mira	217.7	M
RU	15 33.3	−15 20	7.2–14.2	Mira	316.6	M

Double stars

	RA h m	Dec. ° '	PA °	Sep. sec	Mags.
μ	14 49.3	−14 09	355	1.8	5.8, 6.7
α	14 50.9	−16 02	314	231	2.8, 5.2

Globular cluster

M/C	NGC	RA h m	Dec. ° '	Diameter min.	Mag.	
	5897	15 17.4	−21 01	12.6	8.6	H.IV. 19

HYDRA

Brightest stars (on this map)

Star	Proper name	Mag.	Spectrum	RA h m s	Dec. ° ' "
γ Gamma		3.00	G5	13 18 55.2	−23 10 17
π Pi		3.27	K2	14 06 22.2	−26 40 56

Variable star

	RA h m	Dec. ° '	Range (mags.)	Type	Period (d)	Spectrum
R	13 29.7	−23 17	4.0–10.0	Mira	389.6	M

Galaxy

M/C	NGC	RA h m	Dec. ° '	Mag.	Dimensions min.	Type
M83	5236	13 37.0	−29 52	8.2	11.2 × 10.2	Sc

This map shows the small part of Hydra not covered in Maps 29, 30 or 31.

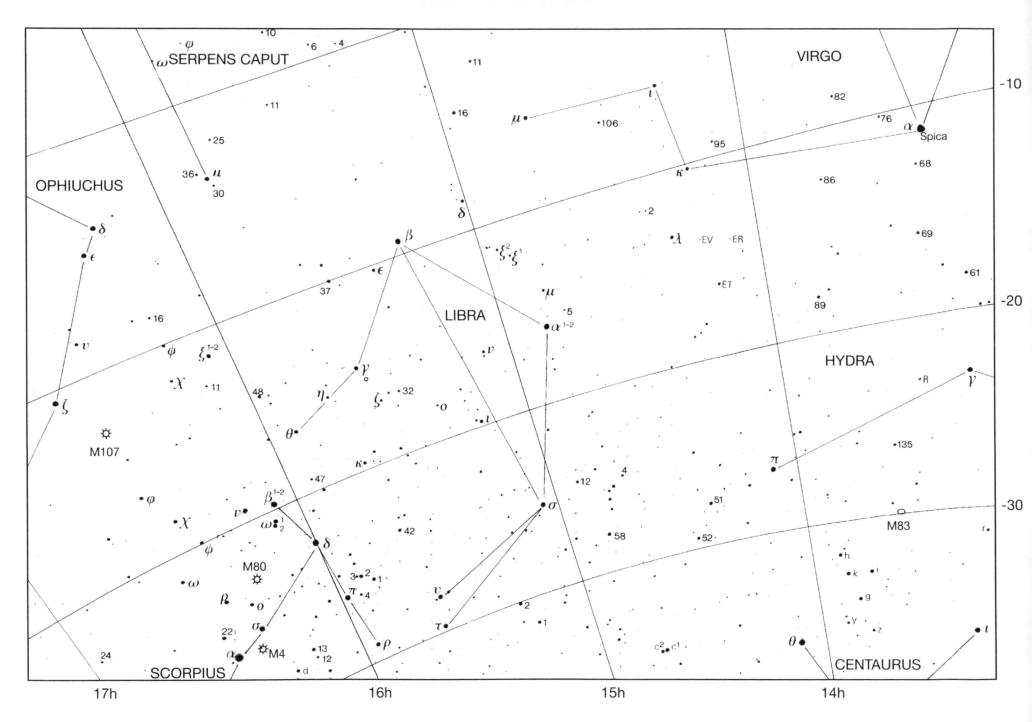

Map 38

SCORPIUS

This map contains the whole of Scorpius, which rivals Orion for the title of the most magnificent constellation in the sky. Moreover, it is one of the very few constellations which really does conjure up at least a vague impression of the object after which it is named; it is not difficult to visualize a scorpion with the long, winding line of bright stars, with the red Antares marking the reptile's heart!

Antares, above the centre of the picture, is a supergiant; it is fiery red – though this does not come out as strongly in the photograph as it does in the sky itself. The very name means "the Rival of Ares" (Mars). Yet Antares is not so remote or powerful as Betelgeux; it is 330 light-years away and 7,500 times as luminous as the Sun. Its diameter is around 200 million miles, greater than that of the Earth's orbit, but the mass is unlikely to be more than ten times that of the Sun. It is very slightly variable, but the fluctuations (magnitude 0.86 to 1.06), unlike those of Betelgeux, are not marked enough to be detectable with the naked eye.

Antares has a 5.4-magnitude companion, which would be a very easy telescopic object if it were not so drowned by the brilliance of the primary; even so, it is not difficult. It makes up a binary system, with a period of 878 years; at present the apparent separation is 2.6 seconds of arc, with a real separation of around 500 astronomical units. The companion – 50 times as luminous as the Sun – is often said to be green, though this may well be due at least partly to a contrast effect. The whole Antares area is enveloped in bright and dark nebulosity, and is well worth sweeping with binoculars or a wide-field telescope.

Antares, like Altair in Aquila, is flanked to either side by fainter stars in this case Sigma (2.9, B1) and Tau (2.8, B0); both are highly luminous, and indeed Sigma is comparable with Antares itself, though it is much further away. It has a companion of magnitude 8.5, at a separation of 20 seconds of arc.

The Scorpion's head is marked by Delta, Beta, Omega and Nu. Beta (2.6) is a fine double, separable with a very small telescope. Nu is a naked-eye double, well shown on the photograph; each component is again double, though the group can in no way compare with the famous quadruple Epsilon Lyræ. Delta (2.3, B0) is the brightest star in the Head.

Epsilon (2.3, K2) below Antares on the photograph is decidedly orange. To its left is the very red H (4.2, M0); compare H with its neighbour N (4.2, B5), which was once thought to be variable, but is now regarded as constant in brightness.

Below Epsilon is the pair made up of Mu¹ (3.0, B1.5) and Mu² (3.6, B2); on the photograph they merge, giving the impression of an elongated star! They share a common motion in space, and are presumably associated, though Mu² is considerably the more distant of the two. Below Mu is an interesting group; there are two brightish stars, Zeta¹ (4.7, B1.5) and Zeta² (3.6, K5), and the bright open cluster NGC 6231 (C76). The two Zetas are not connected. Zeta² is a mere 160 light-years away, but Zeta¹ is a cosmical searchlight; it may well be a member of the cluster, and if so, its luminosity must be 200,000 times that of the Sun, ranking it with Canopus and far outshining Antares. The distance is well over 2,000 light-years. Slightly above and left of C76 is H12, which in

the photograph looks reddish; it is sometimes classed as a separate cluster, but may well be an extension of its neighbour.

The Scorpion's sting is very prominent. It is made up of Iota (3.0, F0) – another immensely remote and luminous star – together with Kappa (2.4, B2); Q (4.3, K0); G (3.2, K2) and the very prominent pair made up of Lambda or Shaula (1.6, B2) and Upsilon or Lesath (2.7, B3). These two look very like a wide naked-eye double, but are not genuinely associated, since Lesath is much the more remote and more powerful; it could match 16,000 Suns.

There are not many bright variables in Scorpius. RR, just above the little clump of stars left of Epsilon, can reach magnitude 5 at its best; at the time of the photograph it was well below maximum, and is only just visible, though its redness makes it identifiable. It is a typical Mira star, and so is RS, below Eta, which was near its maximum of 6.2 at the time of the photograph and is quite prominent. RV, to the left of Epsilon, is a normal Cepheid.

The whole of Scorpius is very rich, and is crossed by the Milky Way. M4, close to Antares, is a fine globular cluster; on the image it appears in the guise of a bluish star, but even binoculars will show it in its true form. It is easy to resolve telescopically, and bears high magnification well. It seems to be only about 7,500 light-years away, in which case it is one of the very nearest of the globular clusters. M80, another globular makes up a triangle with Omicron (4.5, A5) and Rho (3.9, B2); it can be made out as a bluish speck, but is less prominent than M4. In 1850 a bright nova, T Scorpii, flared up in it, and temporarily drowned the cluster itself; unfortunately we do not know a great deal about it. A third globular, NGC 6388, can be seen below Theta (1.9, F0), the southernmost of the bright stars of Scorpius.

Open clusters abound. M6, nicknamed the Butterfly, is easy to find above the Sting. Even brighter is M7, which was recorded by Ptolemy in the second century AD and is a very easy naked-eye object, though it is not so easy to locate on the photograph because it is immersed in a particularly brilliant part of the Milky Way. NGC 6383 and 6416 lie in the same general area. There are two prominent open clusters to the right of Mu on the map, NGC 6242 and 6124 (C75), which looks somewhat bluish. Other clusters are NGC 6322 (near Eta). An interesting but elusive object is the planetary nebula NGC 6302 (C69), nicknamed the Bug Nebula. It can just be seen on the photograph, to the left of Mu, but not easily.

Unquestionably this is one of the finest regions in the whole sky, and of course it contains the brilliant Sagittarius region of the Milky Way, described with Map 43.

SCORPIUS

Brightest stars

Star	Proper name	Mag.	Spectrum	RA h m s	Dec. ° ' "
α Alpha	Antares	0.96v	M1	16 29 24.3	−26 25 55
λ Lambda	Shaula	1.63	B2	17 33 36.4	−37 06 14
θ Theta	Sargas	1.87	F0	17 37 19.0	−42 59 52
ε Epsilon	Wei	2.29	K2	16 50 09.7	−34 17 36
δ Delta	Dschubba	2.32	B0	16 00 19.9	−22 37 18
κ Kappa	Girtab	2.41	B2	17 42 29.0	−39 01 48
β Beta	Graffias	2.64	B0.5+B2	16 05 26.1	−19 48 19
υ Upsilon	Lesath	2.69	B3	17 30 45.6	−37 17 45
τ Tau		2.82	B0	16 35 52.8	−28 12 58
σ Sigma	Alniyat	2.89v	B1	16 21 11.2	−25 35 34
π Pi		2.89	B1	15 58 51.0	−26 06 50
ι¹ Iota¹		3.03	F2	17 47 34.9	−40 07 37
μ¹ Mu¹		3.04	B1.5	16 51 52.1	−38 02 51
G		3.21	K2	17 49 51.3	−37 02 36
η Eta		3.33	F2	17 12 09.0	−43 14 21
μ² Mu²		3.57	B2	16 52 20.0	−38 01 03
ζ² Zeta²		3.62	K5	16 54 34.9	−42 21 41
ρ Rho		3.88	B2	15 56 53.0	−29 12 50
ω¹ Omega¹	Jabhat al Akrab	3.96	B1	16 06 48.3	−20 40 09
ν Nu	Jabbah	4.00	A0+B2	16 11 59.6	−19 27 38

Variable stars

	RA h m	Dec. ° '	Range (mags.)	Type	Period (d)	Spectrum
RT	17 03.5	−36 55	7.0–16.0	Mira	449	M
RR	16 55.6	−30 35	5.0–12.4	Mira	279.4	M
RS	16 56.6	−45 06	6.2–13.0	Mira	320	M
RV	16 58.3	−33 37	6.6–7.5	Cepheid	6.06	F–G
BM	17 41.0	−32 13	6.8–8.7	Semi-reg.	850	K

Double stars

	RA h m	Dec. ° '	PA °	Sep. sec	Mags.	
2	15 53.6	−25 20	274	2.5	4.7, 7.4	
β	16 05.4	−19 48				
11	16 07.6	−12 45	257	3.3	5.6, 9.9	
12	16 12.3	−28 25	073	4.0	5.9, 7.9	
σ	16 21.2	−25 36	273	20.0	2.9, 8.5	
α	16 29.4	−26 26	273	2.7	1.2, 5.4	Binary, 878 y

Open clusters

M/C	NGC	RA h m	Dec. ° '	Diameter min.	Mag.	No of stars
C75	6124	16 25.6	−40 40	29	5.8	100
	6178	16 35.7	−45 38	4	7.2	12
C76	6231	16 54.0	−41 48	15	2.6	
	6242	16 55.6	−39 30	9	6.4	
	6259	17 00.7	−44 40	10	8.0	120
	6281	17 04.8	−37 54	8	5.4	
	6383	17 34.8	−32 34	5	5.5	40 Nebulosity
M6	6405	17 40.1	−32 13	15	4.2	50
(Butterfly Cluster)						
	6416	17 44.4	−32 21	18	5.7	40
M7	6475	17 53.9	−34 49	80	3.3	80

Globular clusters

M/C	NGC	RA h m	Dec. ° '	Diameter min.	Mag.
M80	6093	16 17.0	−22 59	8.9	7.2
M4	6121	16 23.6	−26 32	26.3	5.9
	6388	17 36.3	−44 44	8.7	6.8

Planetary nebulæ

M/C	NGC	RA h m	Dec. ° '	Dimensions sec.	Mag.	Mag. of central star
	6153	16 31.5	−40 15	25	11.5	
C69	6302	17 13.7	−37 06	50	12.8	
(Bug Nebula)						
	6337	17 22.3	−38 29	48		14.7

Objects in Sagittarius and Corona Australis are listed with Map 43; objects in Lupus, Norma with Map 39 and objects in Ara with Map 45.

169

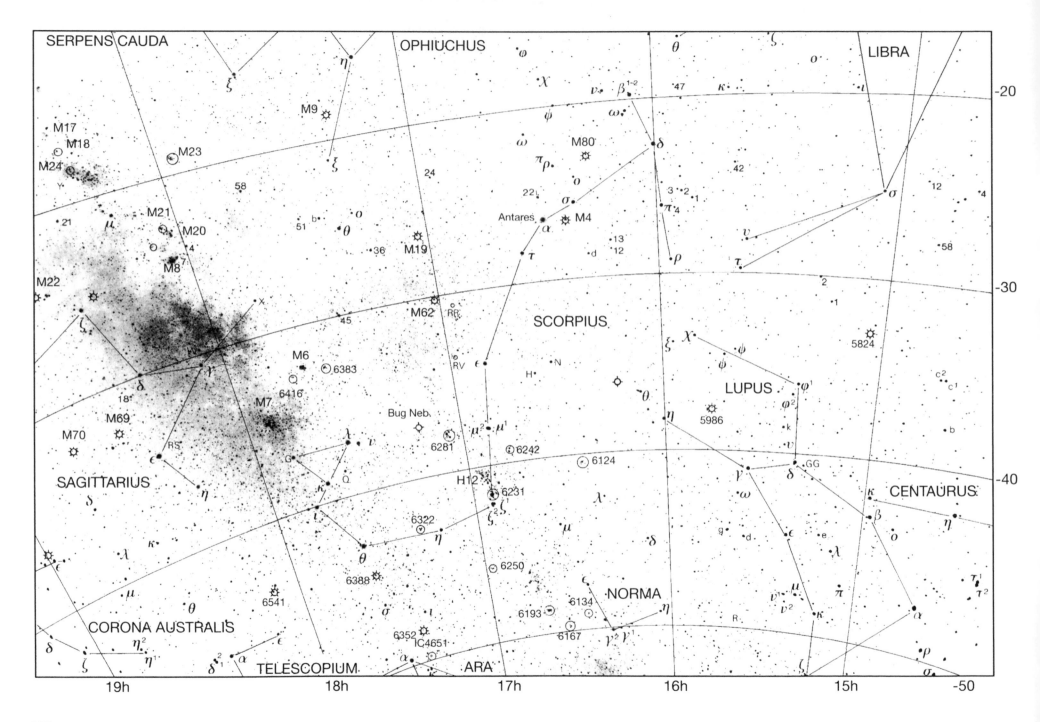

SERPENS CAUDA

OPHIUCHUS

LIBRA

-20

M17
M18
M24
M23

M9

ξ

η

ω

χ

θ

ζ

o

ι

ν β¹⁻²

47

κ

ψ

ω

M80

δ

42

M21
M20
M8

μ

58

51 b

o

θ

24

π ρ

o

σ

3 ·2
1

σ

-30

21

4

36

M19

22

Antares
α

M4

13
12
d

π
4

ρ

ν

58

M22

ζ

δ

18

X

M62
45

RR

τ

SCORPIUS

ε

N

H

θ

ξ χ

ψ

ψ

LUPUS

φ¹
φ²

τ

2

1

5824

c²
c¹

b

-40

M69
M70

SAGITTARIUS

δ

M6
6416

M7

RS

ε

κ

η

G

Q

ι

λ
ν

θ

Bug Neb.

6383
6281

μ²
μ¹

6242
H12

6231
ζ² ζ¹

6124

θ

η

5986

GG

ν

δ

ε

λ

μ

δ

g
d

ε

e

κ
β

η

o

λ

CENTAURUS

κ

σ

λ

κ

μ

θ

6541

CORONA AUSTRALIS

ε

δ
ζ

η²
η¹

δ²
1

α

6388

σ

ι

6250

6352
IC4651

α

TELESCOPIUM

ARA

ε

6193

6134

6167

η

ν² ν¹

NORMA

R

μ

ν¹

ν²

κ

π

α

τ¹
τ²

ρ
ζ

σ

19h

18h

17h

16h

15h

-50

170

Map 39

CENTAURUS, CIRCINUS, LUPUS, NORMA

This map covers a very rich region, including some of the most brilliant parts of the Milky Way; the dark rifts are very evident in the photograph. Almost the whole of Centaurus is shown, together with Lupus, Circinus and Ara; three of the main stars of the Southern Cross appear at the lower right of the image (note the lovely orange hue of Gamma Crucis) and a section of Scorpius is on view at the upper left.

Centaurus is a particularly large and important constellation, containing a variety of interesting objects – and of course the two Pointers to the Cross, Alpha and Beta Centauri, which are shown towards the bottom of the photograph. They give every impression of being true neighbours, but this is not so. Alpha Centauri, at a distance of just over 4 light-years, is the nearest of the bright stars; Beta is 460 light-years away, so that there is absolutely no real connection between the two.

Alpha (which, strangely, has never had a universally-accepted proper name; at various times it has been called Bundula, Toliman and Rigel Kent) is a magnificent binary. This was first noted in 1689, when a Jesuit astronomer, Father Richard, was comet-hunting from Pondicherry in India and happened to turn his telescope towards Alpha Centauri during his search. The brighter component – officially listed, rather confusingly, as Alpha[2] – is a G-type star, 1.7 times more luminous than the Sun; the secondary (Alpha[1]) is orange, of type K, and is less than half as powerful as the Sun, though it is larger. The orbit is decidedly elliptical, and the real separation ranges between 11 and 35 astronomical units, so that when closest together the distance between them is about the same as that between the Sun and Saturn. We see the orbit more or less edge-on, so that the separation and position angle change quite markedly over the years; but it is always an easy double. Few will quarrel with Sir John Herschel's description of it, in 1834, as being "beyond any comparison, the finest double star in the sky".

There is a third member of the system, Proxima, which is slightly closer to us at a distance of 4.3 light-years. It is a feeble red dwarf of spectral type M, showing violent flare activity. As its apparent magnitude is below 15, it is not at all easy to identify even with a large telescope. Its luminosity is a mere 0.00006 that of the Sun.

Beta Centauri, known sometimes as Agena or as Hadar, is a normal B-type star, 10,000 times as powerful as the Sun. (Were it as close to us as Alpha, it would cast shadows.) It has a fourth-magnitude companion at a separation of rather over a second of arc; the brilliance of the primary makes it rather elusive. Undoubtedly the two are genuinely associated, since they share the same motion through space, but they are a very long way apart.

Gamma Centauri (2.2, A0) is a fine though close binary; the components are identical twins. The period is of 84.5 years, so that the position angle and separation change over time; at present the separation is 1 second of arc, but is decreasing, so that for a while in the early twenty-first century Gamma will be a very difficult object. Eta (2.3, B3) and Epsilon (2.3, B1) have faint optical companions, beyond the range of small telescopes.

The other bright star in the constellation is Theta (2.1, K0), which is orange, and is very prominent near the top of the photograph. At a distance of 46 light-years it is one of our nearer neighbours; it is 17 times as luminous as the Sun. It has a proper name, Haratan, which is still used occasionally.

Looking at the photograph, it might appear that there is another brilliant star in Centaurus; Omega, to the right of the centre of the image, and in line with Beta and Epsilon. Yet Omega Centauri is not a single star at all; it is a globular cluster, containing well over a million stars, and with the naked eye it appears as a misty patch of around the fourth magnitude. Owing to the length of exposure the centre of the cluster has become over-exposed. It was known to Ptolemy; Halley, in 1677, first realized that it is a cluster. Its real diameter is over 150 light-years; because of its large apparent size it is probably most spectacular when viewed through binoculars, since it more than fills a normal telescopic field. Its distance from us is of the order of 17,000 light-years.

Another globular, NGC 5286 (C84) lies below and to the left of Omega, directly between Epsilon and Zeta (2.5, B2). It is easy enough to locate; on the photograph it appears as a slightly fuzzy star.

Above and right of Omega in the photograph is NGC 5128 (C77), an active galaxy with a complicated dark lane crossing it. In a small telescope it is a fine sight; on the photograph it looks like a faint star – the integrated magnitude is about 7. It is a strong radio source, and radio catalogues list it as Centaurus A.

There are various open clusters in the Centaur; the brightest of them is NGC 5460, left of and slightly below Zeta, which can be seen with the naked eye in the guise of a slightly enhanced part of the Milky Way. NGC 5617, containing at least 80 stars, is easy to locate, because it lies directly between the Pointers.

Also between the Pointers, slightly above a line joining the two, is the Mira variable R Centauri. It has a rather unusual light-curve; at maximum it may attain magnitude 5.3, and although it was well away from maximum at the time of this photograph its crimson colour is still very marked. V Centauri, forming a triangle with the Pointers, is a classical Cepheid with a range of from magnitude 6.4 to 7.2; the open cluster NGC 5662 lies to its left.

S Centauri, more or less between Gamma and Delta, is an N-type semi-regular with a range of between magnitudes 6 and 7, so that it is always within binocular range. The deep red colour comes out well here; indeed, stars of type N are among the most highly-coloured in the sky.

Above and to the right of Theta, in the northern part of Centaurus, may be seen a triangle of stars of which the brightest is 2 or g (4.2, M1). Just right of the triangle is T, a red semi-regular variable which was here some way below its maximum magnitude of 5.5. A brighter triangle is made up of Nu (4.3, B5), Psi (4.0, A0) and Mu, which is an irregular variable with a range of from magnitude 2.9 to 3.5. Generally it remains at about 3.0, but it is an unstable star, and sometimes produces relatively mild outbursts. Stars which behave in this way are known as Gamma Cassiopeiæ variables, because Gamma Cassiopeiæ, in the far northwest of Cassiopeia (Map 1) is the brightest and best-known of them.

The rest of Centaurus, including the brighter open clusters C97 and C100 (Lambda Centauri) is shown

Continued on page 176

CENTAURUS

Brightest stars

Star	Proper name	Mag.	Spectrum	RA h m s	Dec. ° ' "
α Alpha		−0.27	K1+G2	14 39 36.7	−60 50 02
β Beta	Agena	0.61	B1	14 03 49.4	−60 22 22
θ Theta	Haratan	2.06	K0	14 06 40.9	−36 22 12
γ Gamma	Menkent	2.17	A0	12 41 30.9	−48 57 34
ε Epsilon		2.30	B1	13 39 53.2	−53 27 58
η Eta		2.31v	B3	14 35 30.3	−42 09 28
ζ Zeta	Al Nair al Kentaurus	2.55	B2	13 55 32.3	−47 17 17
δ Delta		2.60	B2	12 08 21.5	−50 43 20
ι Iota		2.75	A2	13 20 35.8	−36 42 44
μ Mu		3.04v	B3	13 49 36.9	−42 28 25
κ Kappa	Ke Kwan	3.13	B2	14 59 09.6	−42 06 15
λ Lambda		3.13	B9	11 35 46.8	−63 01 11
ν Nu		3.41	B2	13 49 30.2	−41 41 16
φ Phi		3.83	B2	13 58 16.2	−42 06 02
τ Tau		3.86	A2	12 37 42.1	−48 32 28
υ¹ Upsilon¹		3.87	B3	13 58 40.7	−44 48 13
d		3.88	G8	13 31 02.6	−39 24 27
π Pi		3.89	B5	11 21 00.4	−54 29 27
σ Sigma		3.91	B3	12 28 02.4	−50 13 51

Variable stars

	RA h m	Dec. ° '	Range (mags.)	Type	Period (d)	Spectrum
X	11 49.2	−41 45	7.0–13.8	Mira	315.1	M
S	12 24.6	−49 26	6.0–7.0	Semi-reg.	65	N
U	12 33.5	−54 40	7.0–14.0	Mira	220.3	M
RV	13 37.5	−56 29	7.0–10.8	Mira	446.0	N
T	13 41.8	−33 36	5.5–9.0	Semi-reg.	60	K–M
μ	13 49.6	−42 28	2.9–3.5	Irreg.		B
R	14 16.6	−59 55	5.3–11.8	Mira	546.2	M
V	14 32.5	−56 53	6.4–7.2	Cepheid	5.49	F–G

Double stars

	RA h m	Dec. ° '	PA °	Sep. sec	Mags.	
γ	12 41.5	−48 58	353	1.4	2.9, 2.9	Binary, 84.5 y
3	13 51.8	−33 00	108	7.9	4.5, 6.0	
β	14 03.8	−60 22	251	1.3	0.7, 3.9	
α	14 39.6	−60 50	215	19.7	0.0, 1.2	Binary, 79.9 y

Open clusters

M/C	NGC	RA h m	Dec. ° '	Diameter min.	Mag.	No of stars
C97	3766	11 36.1	−61 37	12	5.3	100
C100	IC 2944	11 36.6	−63 02	15	4.5	30 λ Cent. cluster
	5138	13 27.3	−59 01	8	7.6	40
	5281	13 46.6	−62 54	5	5.9	40
	5316	13 53.9	−61 52	14	6.0	80
	5460	14 07.6	−48 19	25	5.6	40
	5617	14 29.8	−60 43	10	6.3	80
	5662	14 35.2	−56 33	12	5.5	70

Globular clusters

M/C	NGC	RA h m	Dec. ° '	Diameter min.	Mag.	
C80	5139	13 26.8	−47 29	36.3	3.6	ω Centauri
C84	5286	13 46.4	−51 22	9.1	7.6	

Planetary nebula

M/C	NGC	RA h m	Dec. ° '	Dimensions sec.	Mag.	Mag. of central star
	3918	11 50.3	−57 11	12	8.4	10.9 Blue Planetary

Nebula

M/C	NGC	RA h m	Dec. ° '	Dimensions min.
	5367	13 57.7	−39 59	4.3 Includes IC 4347. Double nucleus

Galaxies

M/C	NGC	RA h m	Dec. ° '	Mag.	Dimensions min.	Type
C83	4945	13 05.4	−49 28	9.5	20.0 × 4.4	SBc
C77	5128	13 25.5	−43 01	7.0	18.2 × 14.3	S0p Cent. A

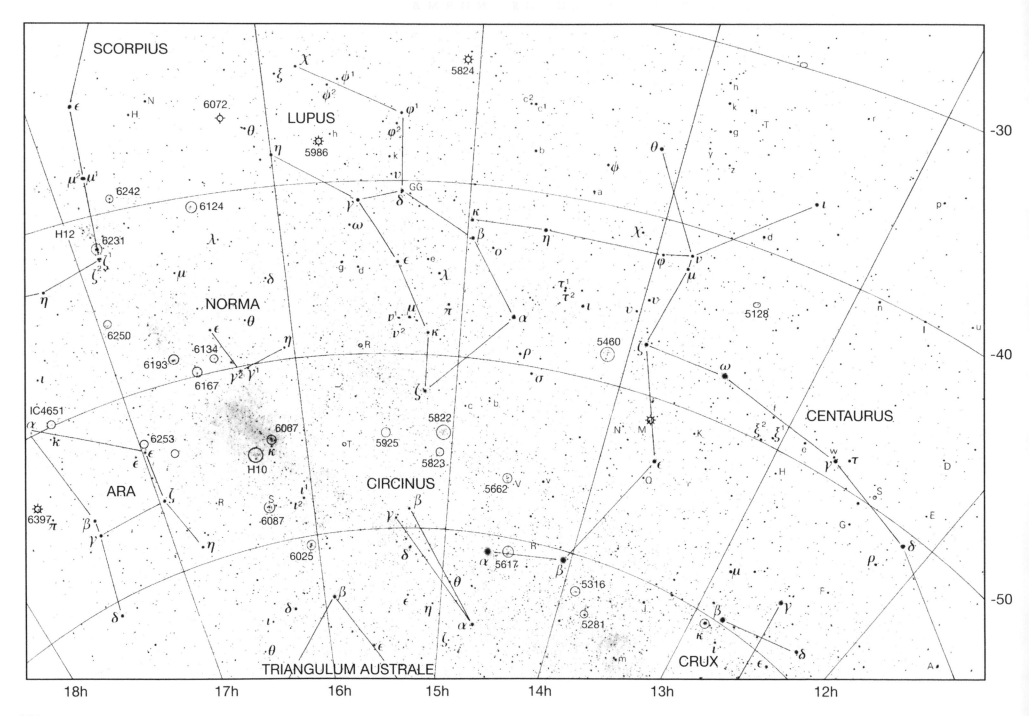

in Map 23, together with the whole of Crux. Note, however, that the dark nebula known as the Coal Sack (C99) is visible here as a dark patch at the extreme bottom of the image.

Circinus is a small constellation close beside the Pointers; its brightest star is Alpha (3.2, F0), which makes up a triangle with Beta (4.1, A3) and Gamma (4.5, B5), the last making up a wide optical pair with a much fainter star; Gamma itself is a very close binary, while Alpha has an 8.6-magnitude companion at a separation of over 15 seconds of arc. Above and to the left of Alpha is Theta Circini, which, like Mu Centauri, is a Gamma Cassiopeiæ-type variable showing occasional outbursts, but the total range is less than half a magnitude.

The open cluster NGC 5823 (C88), above and to the right of Beta Circini, contains at least 100 stars; the integrated magnitude is 8, so that the cluster is easy to find, but care must be taken not to confuse it with the rather similar cluster NGC 5822, which lies above it on the photograph just over the border of Lupus.

Lupus contains over thirty stars above the fifth magnitude, though there is no really distinctive pattern – and although the celestial Wolf is one of Ptolemy's original 48 constellations, there are no well-defined mythological legends attached to it. Several of its stars are double, notably Eta (3.4, B2), Kappa (3.7, B9) and Xi (5.3, A0), all of which are easy telescopic objects. Pi (3.9, B5) and Mu (4.3, B8) are close binaries; Mu also has a third, more distant companion which may be a true member of the group. Epsilon (3.4, B3) is a complex system; the main star, 700 times as luminous as the Sun, is a spectroscopic binary, while the seventh-

magnitude secondary has an orbital period of about 700 years, and there is another star which shares the same motion through space and is presumably associated with the main trio. Just below Gamma (2.8, B3), which is an extremely close and difficult binary, lies Omega Lupi (4.3, M0) whose red colour shows up well. In the north of Lupus, note also the colour contrast between the orange Phi^1 (3.6, K5) and the white Phi^2 (4.5, B3). This little pair is worth looking at with binoculars.

The globular cluster NGC 5986, above and to the left of Gamma and to the right of Eta (3.4, B2) has an integrated magnitude of 7, and is easily found. Another globular, NGC 5824, lies near the top of the photograph; it is less bright than NGC 5986, but is by no means difficult.

In 1006 a brilliant supernova blazed out near Kappa Lupi (3.7, B9) and apparently equalled the quarter-moon, in which case it was the brightest supernova on record. Unfortunately it was not well documented, but its remnant has been traced by radio astronomers. The distance has been given as 3,260 light-years. It seems to have been a Type I outburst, involving the total destruction of the white dwarf component of a binary system.

Just to the left of Theta Lupi (4.3, E3) – one of a trio of dim stars – is the planetary nebula NGC 6072, which is just identifiable; it lies across the boundary of Scorpius. Also identifiable on the map are other open clusters: NGC 6124 (C75), 6231 (C76) and 6242, but these are better shown with Map 38, and are described there.

Norma, added to the sky by La Caille in 1752, contains no bright star – its leader, Gamma^2, is only of the

fourth magnitude – but it is very rich in star-fields and clusters. On the photograph the triangle made up by the three main stars, Gamma (4.0, G8), Eta (4.6, G4) and Epsilon (4.5, B3) appears to the left-hand side. (By some curious quirk, the Greek lettering is incomplete here; Alpha and Beta Normæ do not exist!) The very bright Milky Way region below the Norma triangle contains two interesting open clusters, NGC 6067 (C89) and H10, close to Kappa (4.9, G4). Below these on the photograph is NGC 6087, which contains the Cepheid variable S Normæ. The cluster can be found on the photograph, though on this scale it is naturally impossible to pick out the Cepheid. Several other open clusters lie to the left of the Norma pattern – NGC 6134 and 6167 in Norma itself, and NGC 6193 (C82) and 6250 in Ara. Of these C82 is the most prominent; in the photograph it shows a slightly orange cast.

The Mira variable R Normæ is seen to the right of Eta (4.6, G4), but was well below maximum at the time of the photograph and is not too easy to identify. T Normæ, also a Mira star, lies in a rich region about midway between Eta Normæ and Gamma Circini; here, too, there may be identification problems. Epsilon Normæ is a wide, easy double.

Ara, to the lower left of the photograph, is a fairly distinctive constellation; it is described with Map 45. Part of the Southern Triangle is also shown, including the open cluster NGC 6025 (C95), but the leader of the constellation, Alpha Trianguli Australe (1.9, K2) is just off the bottom of the photograph. Triangulum Australe is described with Map 32.

CIRCINUS

Brightest star

Star	Proper name	Mag.	Spectrum	RA h m s	Dec. ° ' "
α Alpha		3.19	F0	14 42 28.0	−64 58 43

Variable stars

	RA h m	Dec. ° '	Range (mags.)	Type	Period (d)	Spectrum
AX	14 52.6	−63 49	5.6–6.1	Cepheid	5.27	F–G
θ	14 56.7	−62 47	5.0–5.4	Irreg.		B

Open cluster

M/C	NGC	RA h m	Dec. ° '	Diameter min.	Mag.	No of stars
C88	5823	15 05.7	−55 36	10	7.9	100

LUPUS

Brightest stars

Star	Proper name	Mag.	Spectrum	RA h m s	Dec. ° ' "
α Alpha	Men	2.30	B1	14 41 55.7	−47 23 17
β Beta	KeKouan	2.68	B2	14 58 31.8	−43 08 02
γ Gamma		2.78	B3	15 35 08.4	−41 10 00
δ Delta		3.22	B2	15 21 22.2	−40 38 51
ε Epsilon		3.37	B3	15 22 40.7	−44 41 21
ζ Zeta		3.41	G8	15 12 17.0	−52 05 57
η Eta		3.41	B2	16 00 07.1	−38 23 48
φ¹ Phi¹		3.56	K5	15 21 48.3	−36 15 41
κ Kappa		3.72	B9	15 11 56.0	−48 44 16
π Pi		3.89	B5	15 05 07.1	−47 03 04
χ Chi		3.95	B9	15 50 57.4	−33 37 38

Variable star

	RA h m	Dec. ° '	Range (mags.)	Type	Period (d)	Spectrum
GG	15 18.9	−40 47	5.4–6.0	Beta Lyræ	2.16	B+A

Double stars

	RA h m	Dec. ° '	PA °	Sep. sec	Mags.
τ¹	14 26.1	−45 13	204	148.2	4.6, 9.3
π	15 05.1	−47 03	073	1.4	4.6, 4.7
κ	15 11.9	−48 44	144	26.8	3.9, 5.8
ξ	15 56.9	−33 58	049	10.4	5.3, 5.8
η	16 00.1	−38 24	020	15.0	3.6, 7.8

Open clusters

M/C	NGC	RA h m	Dec. ° '	Diameter min.	Mag.	No of stars
	5749	14 48.9	−54 31	8	8.8	30
	5822	15 05.2	−54 31	40	6.5	150

Globular clusters

M/C	NGC	RA h m	Dec. ° '	Diameter min.	Mag.
	5824	15 04.0	−33 04	6.2	9.0
	5927	15 28.0	−50 40	12.0	8.3
	5986	15 46.1	−37 47	9.8	7.1

NORMA

No star above magnitude 4

Variable stars

	RA h m	Dec. ° '	Range (mags.)	Type	Period (d)	Spectrum
R	15 36.0	−49 30	6.5–13.9	Mira	492.7	M
T	15 44.1	−54 59	6.2–13.6	Mira	242.6	M
S	16 18.9	−57 54	6.1–6.8	Cepheid	9.75	F–G

Double star

	RA h m	Dec. ° '	PA °	Sep. sec	Mags.
ε	16 27.2	−47 33	335	22.8	4.8, 7.5

Open clusters

M/C	NGC	RA h m	Dec. ° '	Diameter min.	Mag.	No of stars
	6067	16 13.2	−54 13	13	5.6	100
	6087	16 18.9	−57 54	12	5.4	40
(S Normæ cluster)						
	6134	16 27.7	−49 09	7	7.2	
	6152	16 32.7	−52 37	30	8.1	70

177

Map 40

CYGNUS, LYRA, VULPECULA, SAGITTA

This map covers Cygnus, Lyra, Vulpecula and Sagitta, with portions of Aquila, Delphinus, Lacerta, Cepheus and Hercules. The large triangle formed by the three first-magnitude stars Altair, Deneb and Vega is shown. This was nicknamed "the Summer Triangle" by one of the present authors (PM) in a television programme broadcast long ago, and the name has come into general use, even though it is completely unofficial; the three stars are in different constellations, and the term is quite wrong for the southern hemisphere, where the Triangle is at its best in winter!

The leader of Cygnus, Deneb, is an exceptionally powerful star, around 70,000 times as luminous as the Sun and 1,800 light-years away, so that we now see it as it used to be at the time when Britain was occupied by the Romans. It plays a major rôle in illuminating nearby nebulosity, and binoculars will show the so-called North America Nebula, NGC 7000 (C20); its shape really does recall that of the North American continent. In the same area lie the Pelican Nebula IC 5067, and the Veil Nebula (NGC 6992/5, C33) which is a supernova remnant. These nebulæ do not show up on the photograph, but the Milky Way can be made out, running from the top of the image past Deneb and the Cross of Cygnus.

The Cross is made up of Deneb, Gamma (2.2, F8), Epsilon (2.5, K0), Delta (2.9, A0) and Beta (3.1, K5). The symmetry is rather spoiled by the fact that Beta, often known by its proper name of Albireo, is fainter than the other members of the pattern and is further away from the centre, but to compensate for this Albireo is probably the loveliest coloured double in the entire sky. The primary is golden-yellow, the fifth-magnitude

companion vivid blue. The separation is over 34 seconds of arc, so that the pair may be split with a very small telescope. The primary is a luminous star, 390 light-years away and 700 times as powerful as the Sun; the two make up a very wide binary, but the orbital period must be very long indeed. Delta Cygni has a companion of magnitude 5.3, which is a more difficult telescopic object than might be expected; it, too, is a slow binary.

An inconspicuous but famous double is 61 Cygni, close to Sigma (4.2, B9) and Tau (3.7, F0; Tau is itself a close binary). 61 Cygni was the first star to have its distance measured, by FW Bessel in 1838. It is made up of a pair of red dwarfs separated by over 30 seconds of arc; it has an exceptionally large proper motion of over 5 seconds of arc, which led to its nickname of the "Flying Star" and made Bessel realize that by stellar standards it was probably close. Its distance is 11 light-years, so that it is one of the nearest of the naked-eye stars. Both components have a decidedly orange tint.

Cygnus is rich in variable stars. The most celebrated is Chi, below and slightly to the left of Eta (3.9, K0) on the map; it was in fact the third variable to be recognized as such, following Mira Ceti and Algol in Perseus. Chi Cygni is of spectral type S, and is very red. It has an exceptionally large amplitude; at its best it can rise to magnitude 3.3, brighter than Eta, though at other maxima it does not exceed magnitude 5. At minimum it falls to below 14, and is then hard to identify, particularly since it lies in a crowded region. The period is 407 days, so that it is visible with the naked eye for only a relatively short period in the year. Unfortunately it was faint when this photograph was taken, though it can just be seen to

the left of the line of three stars of which 17 Cygni (5.0, F5) is the brightest.

P Cygni, below and to the left of Gamma on the map, is a variable of very different type. The curved line of stars close to it is a help in identification. It seems to have been first noted in 1600, when it was of the third magnitude and was (wrongly) regarded as a nova. It gradually faded, but in the 1650s it brightened up again to magnitude 3.5. After a further fade, by 1715 it has "settled down" at around magnitude 5, where it has remained ever since. It is very remote, luminous and unstable; it is associated with expanding shells of material, and suffers unpredictable outbursts.

RS Cygni, below Gamma on the map, is a semi-regular variable of type N; its colour makes it easily identifiable. Near Epsilon, the left-hand star of the Cross, are two variables. T Cygni has a K-type spectrum, and fluctuates between magnitudes 5.0 and 5.5; it may be irregular, but its precise classification is uncertain. X Cygni, to its right, is a typical if unremarkable Cepheid.

Two well-known Mira variables are shown in the photograph. R Cygni, close to Theta (4.5, F5) can rise to the sixth magnitude, but at minimum becomes very faint. U Cygni, near Omicron^2 (4.0, K3) has the long period of 462 days, and is of type N, so that it is very red; at the time of the photograph it was well below its maximum of magnitude 5.9, but it can just be seen. Near Rho (4.0, G8) in the northern part of the constellation is the red semi-regular variable W Cygni, seen here not far from its maximum brightness.

There are many clusters and nebulæ in the Swan. M39 is an open cluster near Rho; it is not highly concentrated, but is easy to find. M29, below and to the left of Gamma, is so sparse that it is difficult to see why Messier included in his list; there seems to be little chance of confusing it with a comet! Other clusters are NGC 6811, to the right of Delta on the map, and NGC 6871, between Eta and 28 Cygni (4.9, B3).

Lyra – in mythology, the harp which Apollo gave to the great musician Orpheus – is a small constellation, but with a large number of interesting objects. Of course it is dominated by Vega (0.0, A0), which is 25 light-years away and 52 times as luminous as the Sun. Undoubtedly it is the only brilliant star to have a distinctly bluish tint as seen with the naked eye or through a telescope, though this colour does not show up on the photograph.

Vega has a diameter of about 2.3 million miles. It is of special interest, because in 1983 observations made from the Infra-Red Astronomical Satellite (IRAS) showed that it is associated with a cloud of cool material which may possibly be planet-forming. Vega was the first star found to show this characteristic – and, incidentally, it was also the very first star to be photographed (from Harvard, on 16 July 1850).

Above and to the left of Vega on the map is Epsilon Lyræ, a celebrated multiple star. On the photograph it appears single, but with the naked eye it can be seen to be made up of two components, virtually equal in brightness. Telescopically, each component is seen to be double, so that Epsilon Lyræ is a quadruple system. Delta Lyræ is a double well worth looking at because of the colour contrast; the brighter component is red, while the companion is white. Since they are at about the same

Continued on page 180

CYGNUS

Brightest stars

Star		Proper name	Mag.	Spectrum	RA h m s			Dec. ° ' "		
α	Alpha	Deneb	1.25	A2	20 41 25.8			+45 16 49		
γ	Gamma	Sadr	2.20	F8	20 22 13.5			+40 15 24		
ε	Epsilon	Gienah	2.46	K0	20 46 12.5			+33 58 13		
δ	Delta		2.87	A0	19 44 58.4			+45 07 51		
β	Beta	Albireo	3.08	K5	19 30 43.1			+27 57 35		
ζ	Zeta		3.20	G8	21 12 56.0			+30 13 37		
ξ	Xi		3.72	K5	21 04 55.7			+43 55 40		
τ	Tau		3.72	F0	21 14 47.3			+38 02 44		
κ	Kappa		3.77	K0	19 17 06.0			+53 22 07		
ι	Iota		3.79	A5	19 29 42.1			+51 43 47		
o¹	Omicron¹		3.79	K2	20 13 37.7			+46 44 29		
η	Eta		3.89	K0	19 56 18.2			+35 05 00		
ν	Nu		3.94	A0	20 57 10.2			+41 10 02		
o²	Omicron²		3.98	K3	20 15 28.1			+47 42 51		

Variable stars

	RA h m	Dec. ° '	Range (mags.)	Type	Period (d)	Spectrum
CH	19 24.5	+50 14	6.4–8.7	Z And.	±97	M+B
R	19 36.8	+50 12	6.1–14.2	Mira	426.4	M
RT	19 43.6	+48 47	6.4–12.7	Mira	190.2	M
SU	19 44.8	+29 16	6.5–7.2	Cepheid	3.84	F
χ	19 50.6	+32 55	3.3–14.2	Mira	406.9	S
RS	20 13.4	+38 44	6.5–9.3	Semi-reg.	417	N
P	20 17.8	+38 02	3.0–6.0	Irreg.		Bp
U	20 19.6	+47 54	5.9–12.1	Mira	462.4	N
X	20 43.4	+35 35	5.9–6.9	Cepheid	16.39	F–G
T	20 47.2	+34 22	5.0–5.5	Irreg?		K
W	21 36.0	+45 22	5.0–7.6	Semi-reg.	126	M
SS	21 42.7	+43 35	8.4–12.4	SS Cygni	±50	A–G

Double stars

	RA h m	Dec. ° '	PA °	Sep. sec	Mags.	
β	19 30.7	+27 58	054	34.4	3.1, 5.1	Yellow, blue
δ	19 45.0	+45 07	225	2.4	2.9, 6.3	Binary, 828 y
ψ	19 55.6	+52 26	178	3.2	4.9, 7.4	
γ	20 22.2	+40 15	196	41.2	2.2, 9.9	B is a close dble
61	21 06.9	+38 45	148	30.3	5.2, 6.0	
μ	21 44.1	+28 45	300	1.6	4.8, 6.1	Binary, 507 y

Open clusters

M/C	NGC	RA h m	Dec. ° '	Diameter min.	Mag.	No of stars
	6811	19 38.2	+46 34	13	6.8	70
	6819	19 41.3	+40 11	5	7.3	
	6834	19 52.2	+29 25	5	7.8	50
	6866	20 03.7	+44 00	7	7.6	80
	6871	20 05.9	+35 47	20	5.2	15
	6910	20 23.1	+40 47	8	7.4	50
M29	6913	20 23.9	+38 32	7	6.6	50
	6939	20 31.4	+60 38	8	7.8	80
M39	7092	21 32.2	+48 26	32	4.6	30

Planetary nebula

M/C	NGC	RA h m	Dec. ° '	Dimensions sec.	Mag.	Mag. of central star
	6826	19 44.8	+50 31	30 × 140	9.8	10.4

(Blinking Nebula)

Nebulæ

M/C	NGC	RA h m	Dec. ° '	Dimensions min.	Mag. of illum. star
C27	6888	20 12.0	+38 21	20 × 10	7.4
(Crescent Nebula)					
C34	6960	20 45.7	+30 43	70 × 6	
(Filamentary Neb., 52 Cygni)					
	IC 5067/70	20 50.8	+44 21	80 × 70	
(Pelican Nebula)					
C33	6992/5	20 56.4	+31 43	60 × 8	
(Veil Nebula: SNR)					
C20	7000	20 58.8	+44 20	120 × 100	6
(North America Nebula)					
C19	IC 5146	21 53.5	+47 16	12 × 12	10
(Cocoon Nebula, with sparse cluster)					

distance from us (around 700 light-years), and share the same motion through space, they are presumably associated. Zeta (4.4, A3) is another easy double, and Eta (4.4, B2) has a faint companion.

Beta Lyræ is a famous eclipsing binary. The components, unlike those of Algol, are not very unequal, so that over the full period of just under 13 days there are alternate deep and shallow minima. The range is from magnitude 3.3 to 4.3; in the photograph Beta is practically equal to Gamma (3.2, B9). The real separation between the components is no more than about 45 million kilometres (28 million miles), so that no telescope will show them separately; each must be distorted into the shape of an egg, and since the whole system is surrounded by swirling gas it would indeed be spectacular if seen from close range. Changes in brightness are always going on, and Beta Lyræ has given its name to a whole class of eclipsing binaries. It has a companion of magnitude 8.6.

Two variables in the constellation are worth noting. R Lyræ, well above Vega in the photograph, is a semi-regular with a range of from magnitude 3.9 to 5; good comparison stars are Eta and Theta (each 4.4). In the photograph R was at maximum, and obviously red. Much less evident is RR Lyræ, between Eta Lyræ and Delta Cygni; it is identifiable on the map at about magnitude 7. Its range is only one magnitude, and the period is less than 14 hours. It is a pulsating star, and, like Beta, has given its name to a whole class of similar variables. The importance of RR Lyræ stars lies in the fact that all of them have about the same luminosity – 95 times that of the Sun – so that they can act as "standard candles", and their distances can be found. Many of them occur in stellar clusters, and were once known by the rather misleading name of cluster-Cepheids.

Between Beta and Gamma Lyræ lies the Ring Nebula, M57. This is a planetary nebula – that is to say, an old star which has thrown off its outer layers; telescopically it looks like a tiny, luminous cycle-tyre. It is easy to find, and a small telescope will show it, though the fifteenth-magnitude central star is decidedly elusive. Between Gamma Lyræ and Beta Cygni, to the right of 2 Cygni (4.9, B5) is the globular cluster M56, discovered by Messier in 1779; on the photograph it looks like a star.

Vulpecula, the cosmical Fox, adjoins Cygnus; its brightest star is Alpha (4.4, M0). Much the most interesting object in the constellation is the planetary nebula M27, always known as the Dumbbell because of its shape. It lies to the left of 13 Vulpeculæ (4.6, A0), and is identifiable on the photograph. Its distance is not far short of 1,000 light-years. Vulpecula contains two fairly easily-identified open clusters; NGC 6885 (C37), which forms a triangle with Beta and Eta Cygni, and NGC 6940, below and to the left of Epsilon Cygni, which can admittedly be somewhat elusive because of its low surface brightness.

Sagitta is a small but distinctive constellation – its shape really does resemble that of an arrow! Its leading stars are the orange Gamma (3.5, K5); Delta (3.8, M2); Alpha (4.4, G0) and Beta (also 4.4, G8). Zeta (5.0, A3) close to Delta, is a neat little double; the secondary is itself a very close binary. U Sagittæ, to the right of Delta, is an Algol-type eclipsing system, while S, to the left of Delta, is a normal Cepheid.

Between Delta and Gamma Sagittæ lies the cluster M71, officially classed as a globular, though it is much less condensed than most globulars and has sometimes been listed as a galactic cluster. However, its distance (around 18,000 light-years) seems to indicate that it really is a globular system. On the photograph it appears in the guise of a somewhat bluish star.

LYRA

Brightest stars

Star	Proper name	Mag.	Spectrum	RA h m s	Dec. ° ' "
α Alpha	Vega	0.03	A0	18 36 56.2	+38 47 01
γ Gamma	Sulaphat	3.24	B9	18 58 56.4	+32 41 22
β Beta	Sheliak	3.3 max.	B7	18 50 04.6	+33 21 46
ε Epsilon		3.9 (combined)	A3+A5	18 44 21	+39 38 0

Variable stars

	RA h m	Dec. ° '	Range (mags.)	Type	Period (d)	Spectrum
β	18 50.1	+33 22	3.3–4.3	Beta Lyræ	12.94	B+A
R	18 55.3	+43 57	3.9–5.0	Semi-reg.	46	M
RR	19 25.5	+42 47	7.1–8.1	RR Lyræ	0.57	A–F

Double stars

	RA h m	Dec. ° '	PA °	Sep. sec	Mags.
ε	18 44.3	+39 40	AB+CD 173	207.7	4.7, 5.1
			ε¹ = AB 357	2.6	5.0, 5.1
			ε² = CD 094	2.3	5.2, 5.5
ζ	18 44.8	+37 36	150	43.7	4.3, 5.9
β	18 50.1	+33 22	149	45.7	var., 8.6

Globular cluster

M/C	NGC	RA h m	Dec. ° '	Diameter min.	Mag.
M56	6779	19 16.6	+30 11	7.1	8.2

Planetary nebula

M/C	NGC	RA h m	Dec. ° '	Dimensions sec.	Mag.	Mag. of central star
M57	6720	18 53.6	+33 02	70 × 150	9.7	14.8

(Ring Nebula)

VULPECULA

No star above magnitude 4

Variable stars

	RA h m	Dec. ° '	Range (mags.)	Type	Period (d)	Spectrum
U	19 36.6	+20 20	6.8–7.5	Cepheid	7.99	F–G
T	20 51.5	+28 15	5.4–6.1	Cepheid	4.44	F–G
SV	19 51.5	+27 28	6.7–7.7	Cepheid	45.03	F–K
R	21 04.4	+23 49	7.0–14.3	Mira	136.4	M

Double stars

	RA h m	Dec. ° '	PA °	Sep. sec	Mags.
2 (ES)	19 17.7	+23 02	127	1.8	5.4, 9.2
α-8	19 28.7	+24 40	028	413.7	4.4, 5.8

Open clusters

M/C	NGC	RA h m	Dec. ° '	Diameter min.	Mag.	No of stars
	Cr 399	19 25.4	+20 11	60	3.6	40
(Coat-hanger or Brocchi's Cluster)						
	6823	19 43.1	+23 18	12	7.1	30
	6830	19 51.0	+23 04	12	7.9	20
C37	6885	20 12.0	+26 29	7	5.7	30
	6940	20 34.6	+28 18	31	6.3	60

Planetary nebula

M/C	NGC	RA h m	Dec. ° '	Dimensions sec.	Mag.	Mag. of central star
M27	6853	19 59.6	+22 43	350 × 910	7.6	13.9

(Dumbbell Nebula)

SAGITTA

Brightest stars

Star	Proper name	Mag.	Spectrum	RA h m s	Dec. ° ' "
γ Gamma		3.47	K5	19 58 45.3	+19 29 32
δ Delta		3.82	M2	19 47 23.0	+18 32 03

Variable stars

	RA h m	Dec. ° '	Range (mags.)	Type	Period (d)	Spectrum
U	19 18.8	+19 37	6.6–9.2	Algol	3.38	B–K
S	19 56.0	+16 38	5.3–6.0	Cepheid	8.38	F–G
X	20 05.1	+20 39	7.9–8.4	Semi-reg.	196	N
WZ	20 07.6	+17 42	7.0–15.5	Recurrent nova		Q

(Outbursts 1913, 1946, 1978)

Double star

	RA h m	Dec. ° '	PA °	Sep. sec	Mags.
ζ	19 49.0	+19 09	AB+C 311	8.6	5.5, 8.7
			AB 163	0.3	5.5, 6.2 Binary, 22.8 y

Open cluster

M/C	NGC	RA h m	Dec. ° '	Diameter min.	Mag.	No of stars
	H20	19 53.1	+18 20	7	7.7	15

Globular cluster

M/C	NGC	RA h m	Dec. ° '	Diameter min.	Mag.
M71	6838	19 53.8	+18 47	7.2	8.3

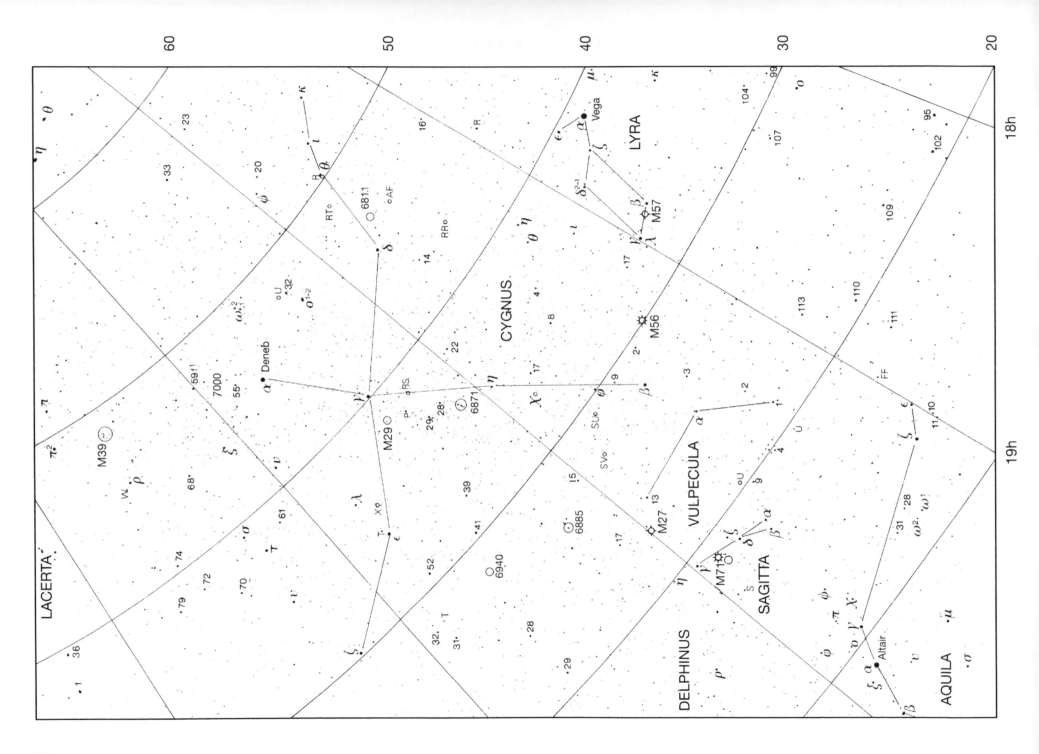

Map 41

LACERTA

This map is centred on the admittedly obscure constellation of Lacerta. Also shown are parts of Pegasus (described in Map 3), Andromeda (Map 2) and Cygnus (Map 40), with a tiny part of Cassiopeia. In Cygnus, note the North America Nebula, near Deneb, and also the Mira variable U, closely left of the little pair made up of Omicron¹ and Omicron²; U Cygni is rather brighter here than it was at the time of the other image, and its colour is more evident. The Andromeda Galaxy, M31, is just on the map, to the extreme left-hand edge.

Lacerta was added to the sky in 1690 by Hevelius; its brightest star is Alpha (3.8, A2), and there is little of immediate interest to the user of a small telescope. The main pattern resembles a distorted diamond pattern, near the top of the photograph; one of its stars, 5 Lacertæ (4.4, M0) is decidedly red. Below it lies 6 Lacertæ (4.5, B2), and to the left of 6 it is possible to identify the red Mira variable R, which was at maximum at the time of this image. Below and to the right of 6 is another Mira star, S, which was well below its best, but can just be seen. The unremarkable open cluster NGC 7243 (C16) lies just to the right of the "diamond".

In 1936 a bright nova, CP Lacertæ, flared up here and reached the second magnitude, though it soon faded back to obscurity. The Milky Way brushes Lacerta, though this is not one of its brighter parts, and it is not very evident on the photograph.

Novæ are not too uncommon, though not very many reach naked-eye visibility. It is now known that a nova is a binary system, made up of a normal star together with a white dwarf companion. The white dwarf pulls material away from its companion, and this material produces an accretion disk round the white dwarf; eventually there is a nuclear outburst in the atmosphere of the dwarf, and gas is ejected at high velocity. At the end of the outburst, the system returns to its normal state.

BRIGHTEST NOVÆ SINCE 1876

Nova	Date	Maximum mag.
Q Cygni	1876	3.0
T Aurigæ	1891	4.2
GK Persei	1901	0.0
DM Geminorum	1903	5.0
DI Lacertæ	1910	4.6
DN Geminorum	1912	3.3
V. 603 Aquilæ	1918	-1.1
V. 476 Cygni	1920	2.0
RR Pictoris	1925	1.1
DQ Herculis	1934	1.2
CP Lacertæ	1936	1.9
BT Monocerotis	1939	4.3
CP Puppis	1942	0.4
V. 533 Herculis	1963	3.2
HR Delphini	1967	3.7
LV Vulpeculæ	1968	4.9
FH Serpentis	1970	4.4
V. 1500 Cygni	1975	1.8
QU Vulpeculæ	1984	5.6
V. 852 Centauri	1986	4.6
Nova Cygni	1992	4.3

LACERTA

Brightest star

Star	Proper name	Mag.	Spectrum	RA h m s	Dec. ° ' "
α Alpha		3.77	A2	22 31 17.3	+50 16 57

Variable stars

	RA h m	Dec. ° '	Range (mags.)	Type	Period (d)	Spectrum
S	22 29.0	+40 19	7.6–13.9	Mira	241.8	M
Z	22 40.9	+56 50	7.9–8.8	Cepheid	10.89	F–G
R	22 43.3	+42 22	8.5–14.8	Mira	299.9	M

Open cluster

M/C	NGC	RA h m	Dec. ° '	Diameter min.	Mag.	No of stars
C16	7243	22 15.3	+49 53	21	6.4	40

NGC 7000, the North America Nebula in the constellation Cygnus (see Map 40)

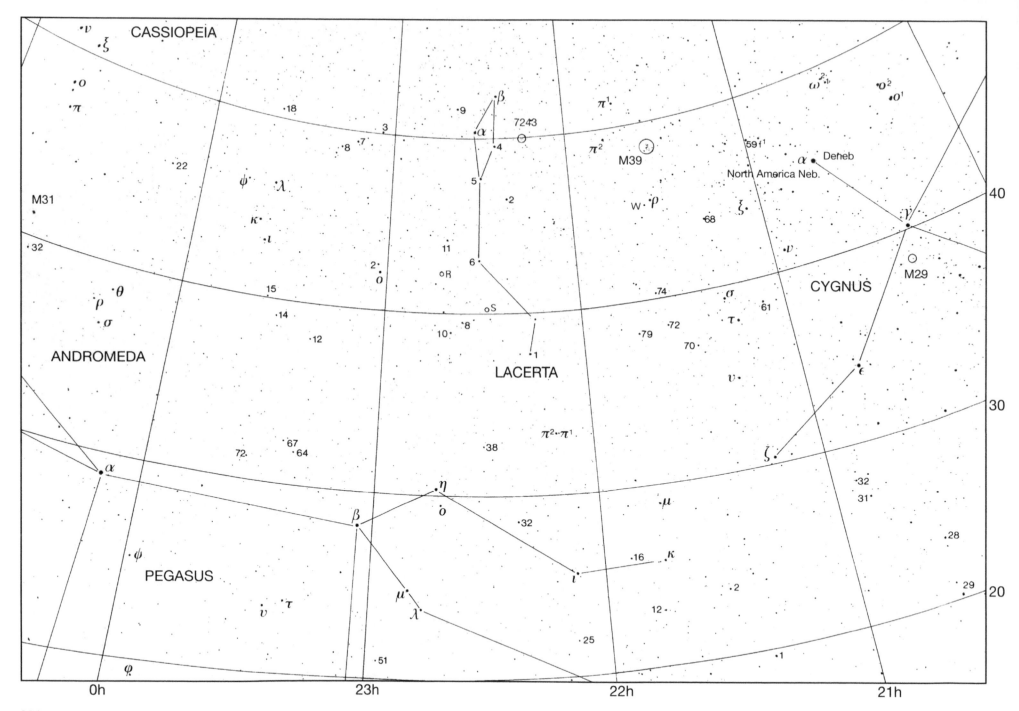

CASSIOPEIA

CYGNUS

LACERTA

ANDROMEDA

PEGASUS

Deneb

North America Neb.

M31

M39

M29

0h   23h   22h   21h

Map 42

AQUILA, SCUTUM, DELPHINUS

This map shows the whole of Aquila, Delphinus, Scutum, Sagitta and Equuleus, with parts of Aquarius and Pegasus. Aquila is one of those constellations which conjures up at least a vague impression of the object after which it is named – in this case an eagle in flight. The leading star, Altair, has a fainter star to either side of it; so too has Antares in Scorpius, but there can be no fear of confusion, since Antares is very red and Altair is pure white. Altair is, incidentally, one of the nearest of the first-magnitude stars; it is only 16.6 light-years away, and a mere 10 times as luminous as the Sun. It is a rapid rotater, and its shape must be distorted into that of an egg.

Gamma Aquilæ, often still known by its proper name of Tarazed, is of type K, and obviously orange; it is much more powerful than Altair, and could match 700 Suns, but it is of course much further away (186 light-years). Beta Aquilæ, a magnitude fainter than Gamma, is of type G8 and only $4^{1}/_{2}$ times as luminous as the Sun; its distance is 36 light-years.

Probably the most important object in the Eagle is Eta Aquilæ, a classical Cepheid variable whose fluctuations were discovered at about the same time as those of Delta Cephei itself. The range is from magnitude 3.5 to 4.4, so that Theta (3.3, B9) and Delta (3.4, F0) make good comparisons for it. The Mira variable R Aquilæ, to the right of Altair on the map, can reach 5.5 magnitude when at its best. The open cluster NGC 6755, slightly above right of Delta, has an integrated magnitude of 7.5. To its right is Theta Serpentis, a very wide, easy double star which can be split with a very small telescope; its components are virtual twins.

V Aquilæ, close to Lambda (3.4, B8.5) is a semi-regular variable with a spectrum of type N, so that it is very red. Adjoining the southernmost part of Aquila is the little constellation of Scutum, which has only one star above the fourth magnitude – Alpha (3.8, K3) (not shown here) – but contains the famous open cluster M11, nicknamed the "Wild Duck". It is fan-shaped, and a glorious sight in even a modest telescope. M26, near Delta (4.7, F3) is a less prominent open cluster.

Delphinus is a compact little constellation – unwary observers have even been known to confuse it with the Pleiades! Its brightest stars are Beta (3.5, F5) and Alpha (3.8, B9). Both have curious proper names – Svalocin and Rotanev (they were christened by an astronomer named Nicolaus Venator). Gamma is an easy double.

Delphinus contains two prominent red variables, U and EU, both of which are easy to follow with binoculars. In 1967, a nova, HR, flared up near these two, and reached magnitude 3.7; it remained visible with the naked eye for months, but has now faded back to its pre-outburst magnitude of 14.

The Milky Way flows through this whole area, and there are many rich fields.

R Scuti is shown on the map, but is not too easy to identify because it is in a very rich region. It lies between Beta Scuti (4.2, G5) and the cluster M11. It is the brightest of the RV Tauri-type variables, which are pulsating supergiants; there are alternate deep and shallow minima, but there are also marked irregularities. At its best, R Scuti can reach magnitude 4.4; at minimum it can drop to 8.2. The mean period is 140 days.

AQUILA

Brightest stars

Star	Proper name	Mag.	Spectrum	RA h m s	Dec. ° ' "
α Alpha	Altair	0.77	A7	19 50 46.8	+08 52 06
γ Gamma	Tarazed	2.72	K3	19 46 15.4	+10 36 48
ζ Zeta	Dheneb	2.99	B9	19 05 24.4	+13 51 48
θ Theta		3.23	B9	20 11 18.1	−00 49 17
δ Delta		3.36	F0	19 25 29.7	+03 06 53
λ Lambda	Althalimain	3.44	B8.5	19 06 14.7	−04 52 57
η Eta		3.5 max.	G0v	19 52 28.1	+01 00 20
β Beta	Alshain	3.71	G8	19 55 18.5	+06 24 24

Variable stars

	RA h m	Dec. ° '	Range (mags.)	Type	Period (d)	Spectrum
V	19 04.4	−05 41	6.6–8.4	Semi-reg.	353	N
R	19 06.4	+08 14	5.5–12.0	Mira	284.2	M
TT	19 08.2	+01 18	6.4–7.7	Cepheid	13.75	F–G
W	19 15.4	−07 03	7.3–14.3	Mira	490.4	S
U	19 29.4	−07 03	6.1–6.9	Cepheid	7.02	F–G
η	19 52.5	+01 00	3.5–4.4	Cepheid	7.18	F–G

Double stars

	RA h m	Dec. ° '	PA °	Sep. sec	Mags.
31	19 25.0	+11 57	343	105.6	5.2, 8.7
ν	19 26.5	+00 20	288	201.0	4.7, 8.9
γ	19 46.3	+10 37	258	132.6	2.7, 10.7
α	19 50.8	+08 52	301	165.2	0.8, 9.5
57	19 54.6	−08 14	170	35.7	5.8, 6.5

Open clusters

M/C	NGC	RA h m	Dec. ° '	Diameter min.	Mag.	No of stars
	6709	18 51.5	+10 21	13	6.7	40
	6755	19 07.8	+04 14	15	7.5	100

SCUTUM

Brightest star

Star	Proper name	Mag.	Spectrum	RA h m s	Dec. ° ' "
α Alpha		3.85	K3	18 35 12.1	−08 14 39

Variable stars

	RA h m	Dec. ° '	Range (mags.)	Type	Period (d)	Spectrum
R	18 47.5	−05 42	4.4–8.2	RV Tauri	140	G–K
S	18 50.3	−07 54	7.0–8.0	Semi-reg.	148	N

Open clusters

M/C	NGC	RA h m	Dec. ° '	Diameter min.	Mag.	No of stars
	6664	18 36.7	−08 13	16	7.8	50
(EV Scuti Cluster)						
M26	6694	18 45.2	−09 24	15	8.0	30
M11	6705	18 51.1	−06 16	14	5.8	500
(Wild Duck Cluster)						

Globular cluster

M/C	NGC	RA h m	Dec. ° '	Diameter min.	Mag.
	6712	18 53.1	−08 42	7.2	8.2

Nebula

M/C	NGC	RA h m	Dec. ° '	Dimensions min.	Mag. of illum. star
	IC 1287	18 31.3	−10 50	4.4 × 3.4	5.5

DELPHINUS

Brightest stars

Star	Proper name	Mag.	Spectrum	RA h m s	Dec. ° ' "
β Beta	Rotanev	3.54	F5	20 37 32.8	+14 35 43
α Alpha	Svalocin	3.77	B9	20 39 38.1	+15 54 43
γ Gamma		3.9	G5+F8	20 46 39.3	+16 07 27

Variable stars

	RA h m	Dec. ° '	Range (mags.)	Type	Period (d)	Spectrum
R	20 14.9	+09 05	7.6–13.8	Mira	284.9	M
EU	20 37.9	+18 16	5.8–6.9	Semi-reg.	59	M
U	20 45.5	+18 05	5.7–7.6	Semi-reg.	110	M

Double star

	RA h m	Dec. ° '	PA °	Sep. sec	Mags.
γ	20 46.7	+16 07	268	9.6	4.5, 5.5

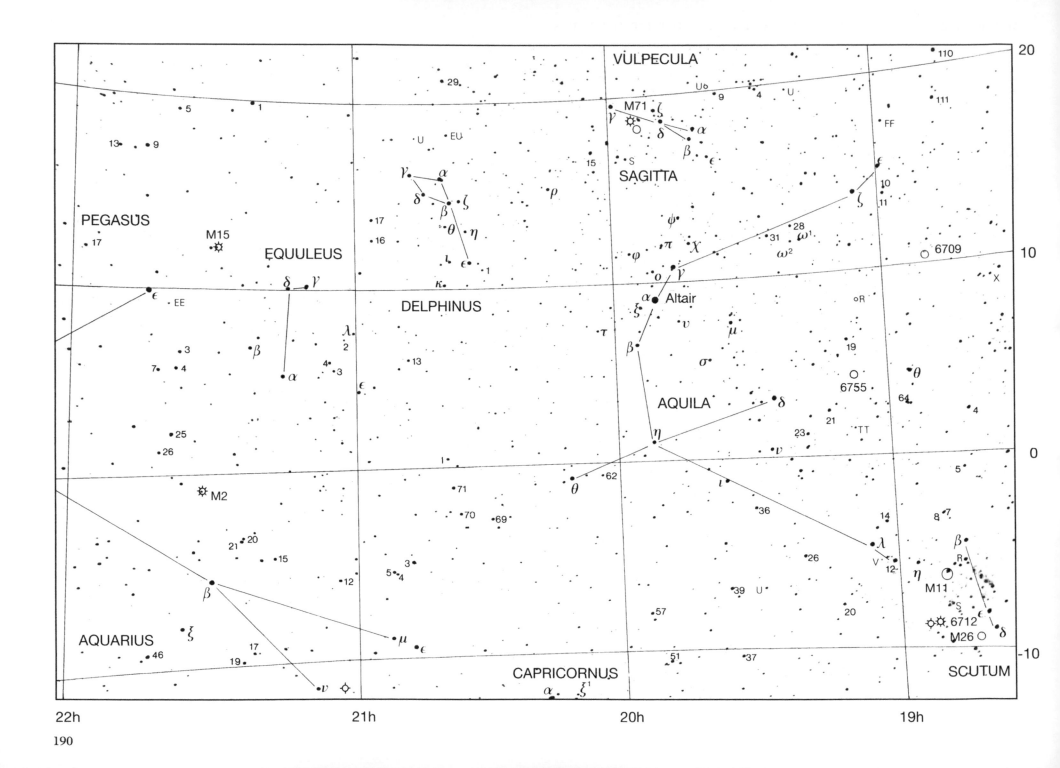

Map 43

SAGITTARIUS, CORONA AUSTRALIS

This map covers Sagittarius and Corona Australis, with the glorious star-clouds which mask our view of that mysterious region, the centre of the Galaxy. Parts of Capricornus, Aquila, Scutum, Ophiuchus, Scorpius, Ara, Telescopium and Microscopium are also shown.

The brilliant object at the top right of the photograph is Jupiter. Its magnitude was then −2.3, and it is very evident that it far outshines any star – though in fact there are no first-magnitude stars on this map.

Sagittarius has no really distinctive shape (though it is often likened to a teapot!). The brightest star is Epsilon (1.8, B9), slightly to the right of the centre of the photograph; below it and to the right is the reddish Eta (3.1, M3), which has an optical companion of magnitude 7.8. Above Epsilon on the map are Delta (2.7, K2) and Gamma (3.0, K0), both slightly orange; their distances are respectively 81 and 117 light-years, so that they are far closer than the Milky Way stars in the background. Immediately above Delta and Gamma on the image is a particularly rich part of the Milky Way, well worth sweeping with binoculars or a wide-field telescope; there is bright nebulosity, and also dark, winding rifts indicating obscuring material. The area contains the Cepheid variable W Sagittarii, which was then just below the fourth magnitude.

The second brightest star in Sagittarius is Sigma (2.0, B3), so that the sequence of Greek letters has been disregarded; Alpha Sagittarii, near the bottom of the picture, is only of the fourth magnitude.

To the right of Sigma are Phi (3.2, B8) and Lambda (2.8, K2). Well above Lambda, near a triangle of faint stars, is Mu (3.8, B8); it looks ordinary enough, but is a

very powerful supergiant 60,000 times as luminous as the Sun, so that it is the equal of Rigel in Orion. It is 1,200 light-years away, and very slightly variable. Zeta (2.6, A2) is a binary with almost equal components, but the separation is only 0.3 of a second of arc, so that this is a difficult test. The orbital period is 21 years.

Variable stars abound in Sagittarius. Near the bottom of the photograph, close to the red Iota (4.1, K0) is a Mira star, RU; it can rise to magnitude 6, but was well on the way to minimum when this photograph was taken. RR, another Mira star lies to the left-hand side of the photograph, below the triangle of stars of which Omega (4.7, G5) is the brightest; it is here seen near its maximum, magnitude 5.6, and its redness is very evident – compare it with the non-variable 62 Sagittarii (4.6, M4) to its left. X Sagittarii, above and right of Gamma, is a normal Cepheid with a small range, magnitude 4.2 to 4.8, and another Cepheid, Y, is identifiable above and left of Mu. The magnitude here is between 5 and 6.

RY Sagittarii, below Zeta, is of different type; it is an R Coronæ Borealis star, remaining at maximum for most of the time, but with sudden falls to minimum. At the time of the photograph it was in its "rest" state, magnitude 6, and is on the fringe of naked-eye visibility; when it fades it falls to below magnitude 15, beyond the range of small telescopes. Minima can never be predicted; it depends upon how much soot builds up in the star's atmosphere.

It is no surprise to find that the Archer is very rich in clusters and nebulæ. One of the most spectacular of these objects is M8, the Lagoon Nebula, which also contains a galactic cluster (NGC 6520). It lies below

and to the right of Mu, and on the photograph appears very red, though this colour is not evident when the nebula is viewed through a telescope. The Lagoon was recorded by Flamsteed as long ago as 1680, and is visible with the naked eye; it is so large that the best views are obtained with a wide-angle telescope eyepiece. Closely above it is the Trifid Nebula, M20, so called because of the dark lanes seen in it telescopically. The real diameter is of the order of 20 light-years, and, like the Lagoon, it is around 6,500 light-years away.

Also nearby are two open clusters, M21 and NGC 6546, both of which are easy to locate but are not remarkable.

Another famous nebula is M17, variously nicknamed the Omega, the Horseshoe and the Swan. On the photograph it appears as a deep-red spot near the top centre; the best guide star to it is Gamma Scuti (4.7, A2). Here, too, we have bright nebulosity and dark obscuring matter. Just below it on the image is the open cluster M18, also easy to identify; Messier himself discovered it, in 1764. M25, to the left of Mu, is quite conspicuous, and in the photograph looks slightly greenish, though again this colour is not evident visually.

M22, near Lambda, was the first globular cluster to be discovered – in 1665 – by an otherwise obscure astronomer named Abraham Ihle. It is easy to find, and is a fine object, slightly non-circular in outline and relatively loose by globular standards, so that the outer parts at least can be resolved with a modest telescope. It has been compared with Omega Centauri, though it is rather smaller and two magnitudes fainter. Most of the other globular clusters in Sagittarius appear more or less stellar *Continued on page 196*

SAGITTARIUS

Brightest stars

Star		Proper name	Mag.	Spectrum	RA h m s	Dec. ° ' "
ε	Epsilon	Kaus Australis	1.85	B9	18 24 10.2	−34 23 05
σ	Sigma	Nunki	2.02	B3	18 55 15.7	−26 17 48
ζ	Zeta	Ascella	2.59	A2	19 02 36.5	−29 52 49
δ	Delta	Kaus Meridionalis	2.70	K2	18 20 59.5	−29 49 42
λ	Lambda	Kaus Borealis	2.81	K2	18 27 58.1	−25 25 18
π	Pi	Albaldah	2.89	F2	19 09 45.6	−21 01 25
γ	Gamma	Alnasr	2.99	K0	18 05 48.3	−30 25 26
η	Eta		3.11	M3	18 17 37.5	−36 45 42
φ	Phi		3.17	B8	18 45 39.2	−26 59 27
τ	Tau		3.32	K1	19 06 56.2	−27 40 13
ξ²	Xi²		3.51	K1	18 57 43.6	−21 06 24
o	Omicron		3.77	gG8	19 04 40.8	−21 44 30
μ	Mu	Polis	3.86v	B8	18 13 45.6	−21 03 32
ρ¹	Rho¹		3.93	F0	19 21 40.2	−17 50 50
β¹	Beta¹	Arkab	3.93	B8	19 22 38.1	−44 27 32
α	Alpha	Rukbat	3.97	B8	19 23 53.0	−40 36 58

Variable stars

	RA h m	Dec. ° '	Range (mags.)	Type	Period (d)	Spectrum
X	17 47.6	−27 50	4.2–4.8	Cepheid	7.01	F
W	18 05.0	−29 35	4.3–5.1	Cepheid	7.59	F–G
VX	18 08.1	−22 13	6.5–12.5	Semi-reg.	732	M
RS	18 17.6	−34 06	6.0–6.9	Algol	2.41	B+A
Y	18 21.4	−18 52	5.4–6.1	Cepheid	5.77	F
U	18 31.9	−19 07	6.3–7.1	Cepheid	6.74	F–G
YZ	18 49.5	−16 43	7.0–7.7	Cepheid	9.55	F–G
RY	19 16.5	−33 31	6.0–15.0	R Coronæ		Gp
R	19 16.7	−19 18	6.7–12.8	Mira	268.8	M
AQ	19 34.3	−16 22	6.6–7.7	Semi-reg.	200	N
RR	19 55.9	−29 11	5.6–14.0	Mira	334.6	M
RU	19 58.7	−41 51	6.0–13.8	Mira	240.3	M
RT	20 17.7	−39 07	6.0–14.1	Mira	305.3	M

Double stars

	RA h m	Dec. ° '	PA °	Sep. sec	Mags.	
21	18 25.3	−20 32	289	1.8	4.9, 7.4	
ζ	19 02.6	−29 53	320	0.3	3.3, 3.4	Binary, 21.2 y
η	18 17.6	−36 46	105	3.6	3.2, 7.8	
β¹	19 22.6	−44 28	077	28.3	3.9, 8.0	

(Wide naked-eye pair with β²)

Open clusters

M/C	NGC	RA h m	Dec. ° '	Diameter min.	Mag.	No of stars
M23	6494	17 56.8	−19 01	27	5.5	150
	6520	18 03.4	−27 54	6	7.6	60
(In M20)						
M21	6531	18 04.6	−22 30	13	5.9	70
	6530	18 04.8	−24 20	15	4.6	
(In M20)						
M24		18 16.9	−18 29	90	4.5	
(Star-cloud; not a true cluster)						
M18	6613	18 19.9	−17 08	9	6.9	20
M25	IC 4725	18 31.6	−19 15	32	4.6	30
(υ Sagittarii Cluster)						
	6716	18 54.6	−19 53	7	6.9	20

Globular cluster

M/C	NGC	RA h m	Dec. ° '	Diameter min.	Mag.
M28	6626	18 24.5	−24 52	11.2	6.9
M69	6637	18 31.4	−32 21	7.1	7.7
M22	6656	18 36.4	−23 54	24.0	5.1
M54	6715	18 55.1	−30 29	9.1	7.7
M70	6681	18 43.2	−32 18	7.8	8.1
M55	6809	19 40.0	−30 58	19.0	6.9
M75	6864	20 06.1	−21 55	6.0	8.6

Planetary nebula

M/C	NGC	RA h m	Dec. ° '	Dimensions sec.	Mag.	Mag. of central star
	6818	19 44.0	−14 09	17	9.9	13.0

Nebulæ

M/C	NGC	RA h m	Dec. ° '	Dimensions min.	Mag. of illum. star
M20	6514	18 02.6	−23 02	29 × 27	7.6
(Trifid Nebula)					
M8	6523	18 03.8	−24 23	90 × 40	6.0
(Lagoon Nebula)					
M17	6618	18 20.8	−16 11	46 × 37	7.0
(Omega Nebula)					

CORONA AUSTRALIS

No star above magnitude 4

Double stars

	RA h m	Dec. ° '	PA °	Sep. sec	Mags.	
κ	18 33.4	−38 44	359	21.6	5.9, 5.9	
λ	18 43.8	−38 19	214	29.2	5.1, 9.7	
γ	19 06.4	−37 04	109	1.3	4.8, 5.1	Binary, 120.4 y

Globular cluster

M/C	NGC	RA h m	Dec. ° '	Diameter min.	Mag.
C78	6541	18 08.0	−43 42	13.1	6.6

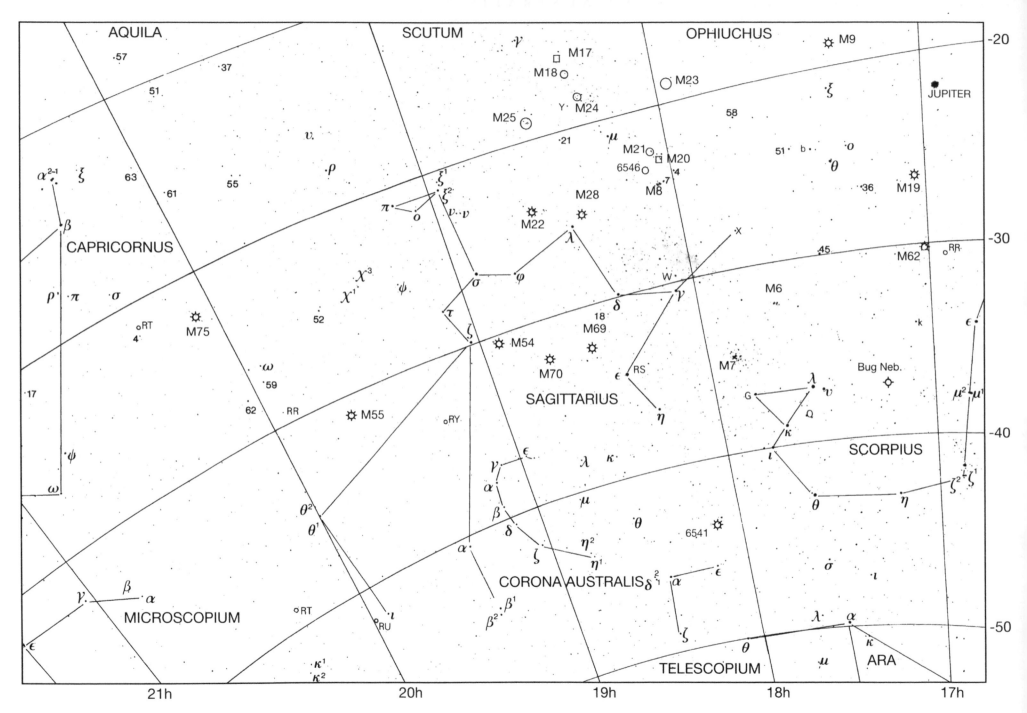

AQUILA   SCUTUM   OPHIUCHUS

-20

-30

-40

-50

CAPRICORNUS

SAGITTARIUS

SCORPIUS

MICROSCOPIUM

CORONA AUSTRALIS

TELESCOPIUM   ARA

JUPITER

Bug Neb.

21h   20h   19h   18h   17h

Map 44

CAPRICORNUS

The whole of Capricornus is included here, together with a large part of Aquarius and portions of Aquila, Piscis Australis and Sagittarius. All in all it is rather a barren area, but at the time of the photograph Saturn was in Capricornus, and is shown just below the middle of the picture. Its magnitude was less than 0.3. There are no really bright stars anywhere on this map.

The celestial Sea-Goat contains little of immediate interest to the telescope-user. The brightest star is Delta (2.9, A5), 49 light-years away and 13 times as luminous as the Sun. Alpha is a very wide double, separable with the naked eye. The magnitudes of the components are 4.2, and 3.6; the fainter – known, rather confusingly, as Alpha[1] – is 1,600 light-years away, while the distance of the brighter member, Alpha[2], is only 117 light-years, so that there is absolutely no real connection between them. On the photograph it is just possible to separate them. Beta (3.1, F8) is an easy telescopic double; the companion is of the sixth magnitude. The two components have a common motion in space, but are at least 9,500 astronomical units apart.

It is interesting to remember that in 1846 the planet Neptune was discovered only about four degrees from Delta Capricorni. Its position had been worked out by the French astronomer UJJ Le Verrier, and the actual identification was made by Johann Galle and Heinrich D'Arrest at Berlin, using the Fraunhofer telescope there. (Similar calculations had been made in England by John Couch Adams, but these were not known to the German observers.)

The semi-regular variable RT Capricorni can be seen below Sigma (5.5, K4). Though it was well below its maximum of magnitude 6.5 when this photograph was taken, it is still identifiable because of its intensely red colour. RS Capricorni, just above and slightly to the left of Theta (4.1, A0) is also identifiable because of its colour.

The globular cluster M30, discovered by Messier himself in 1764, is reasonably bright, though not too easy to identify on the photograph; it lies left and slightly below Zeta (3.7, G4), very close to the star 41 Capricorni (5.3, G9). It is around 41,000 light-years away, and has a real diameter of about 75 light-years. A small telescope shows it easily.

Aquarius was shown in Map 6, and the interesting objects are listed there, but there were two Messier objects which were just outside the borders of that map. Both are close to the Saturn Nebula, NGC 7009 (C55); M72 is a globular cluster discovered by Méchain in 1780; it is more open than M2. Adjoining it is M73. This was one of Messier's mistakes; M73 is not a cluster at all, or even a rich area. It is made up of a few faint stars, by no means easy to identify either on the photograph or in the actual sky!

On this map, note the so-called "Water Jar" of Aquarius, made up of Gamma, Eta, Pi and the fine binary Zeta. It is interesting to recall that Zeta Aquarii was one of the first binaries to be identified as such – by William Herschel, in 1804.

A small part of Sagittarius is shown here, including the globular cluster M75. To the top right appears part of Aquila. Eta Aquilæ is a Cepheid variable, identified by Pigott in 1784 at about the same time that John Goodricke discovered the variability of Delta Cephei

in the photograph, but inspection with a telescope (or, in some cases, binoculars) reveals their true nature: M28 (near Lambda), M55 (below and left of Zeta), M69 (below left of Delta – note the little chain of stars below it). M70 (forming a triangle with Zeta and Delta) and M75 (well to the left of the main pattern, not far from the border of Capricornus; it tends to be elusive both on the photograph and in the sky, because it lies in a relatively barren area).

It is worth identifying M24, easily seen with the naked eye between Mu Sagittarii and Gamma Scuti. Even though it was listed by Messier, who first noted it in 1764, it is not a true cluster or nebula; it is nothing more than a bright part of the Milky Way, with no claim to separate identity. On the photograph it is very prominent.

Corona Australis, the Southern Crown (sometimes called Corona Austrinus) is not nearly so conspicuous as its northern namesake, and its brightest stars, Alpha and Beta, are only of magnitude 4.1. However, the curved line of stars, above and right of Alpha Sagittarii, is distinctive enough. Eta and Epsilon are close binaries. Well to the right of the semi-circle, beyond Theta (4.6, G5) is the globular cluster NGC 6541 (C78), which can be seen on the photograph as a faint point; it is not a difficult object. Kappa is an easy double, Gamma a close binary.

The Scorpion's Sting appears at the right-hand side of the image; note the very prominent pair consisting of Lambda and Upsilon, and also the fine open clusters M7 (which looks bluish here) and M6 (the Butterfly). The Bug Nebula, NGC 6302 (C69) can just be made out right of the Lambda-Upsilon pair. Scorpius is described with Map 38.

itself; if Pigott's announcement had come first, our Cepheids might well have become known as Aquilids! A good comparison star is Theta Aquilæ (3.2, B9). In this photograph the two appear almost equal, though the official maximum magnitude of Eta is only 3.5. At minimum it drops to 4.4, equal to Iota (4.4, B5), which is shown very near the right-hand edge of the photograph. The whole of Aquila is shown in Map 42, and the interesting objects are listed there.

CAPRICORNUS

Brightest stars

Star	Proper name	Mag.	Spectrum	RA h m s	Dec. ° ' "
δ Delta	Deneb al Giedi	2.87v	A5	21 47 02.3	−16 07 38
β Beta	Dabih	3.08	F8	20 21 00.5	−14 46 53
α² Alpha²	Al Giedi	3.57	G9	20 18 03.1	−12 32 42
γ Gamma	Nashira	3.68	F0	21 40 05.2	−16 39 45
ζ Zeta	Yen	3.74	G4	21 26 39.9	−22 24 41

Variable stars

	RA h m	Dec. ° '	Range (mags.)	Type	Period (d)	Spectrum
RT	20 17.1	−21 19	6.5–8.1	Semi-reg.	393	N
RS	21 07.2	−16 25	7.0–9.0	Semi-reg.	340	M

Double stars

	RA h m	Dec. ° '	PA °	Sep. sec	Mags.	
α	20 18.1	−12 33	291	377.7	3.6, 4.2	(α¹–α²)
α¹	20 17.6	−12 30	AB 182	44.3	4.2, 13.7	
			AC 221	45.4	9.2	
α¹	20 18.1	−12 33	AB 172	6.6	3.6, 11.0	
			AD 156	154.6	9.3	
			BC 240	1.2	11.3	
σ	20 19.6	−19 07	179	55.9	5.5, 9.0	
β	20 21.0	−14 47	267	205	3.1, 6.0	B is double
π	20 27.3	−18 13	148	3.2	5.3, 8.9	

Globular cluster

M/C	NGC	RA h m	Dec. ° '	Diameter min.	Mag.
M30	7099	21 40.4	−23 11	11.0	7.5

AQUARIUS

Asterism

M/C	NGC	RA h m	Dec. ° '	
M73	6994	20 58.9	−12 38	Four stars; not a true cluster

Globular cluster

M/C	NGC	RA h m	Dec. ° '	Diameter min.	Mag.
M72	6981	20 53.5	−12 32	5.9	9.3

Aquarius is shown in Map 6, and interesting objects are listed there, but M72 and M73 were just off the border of that map.

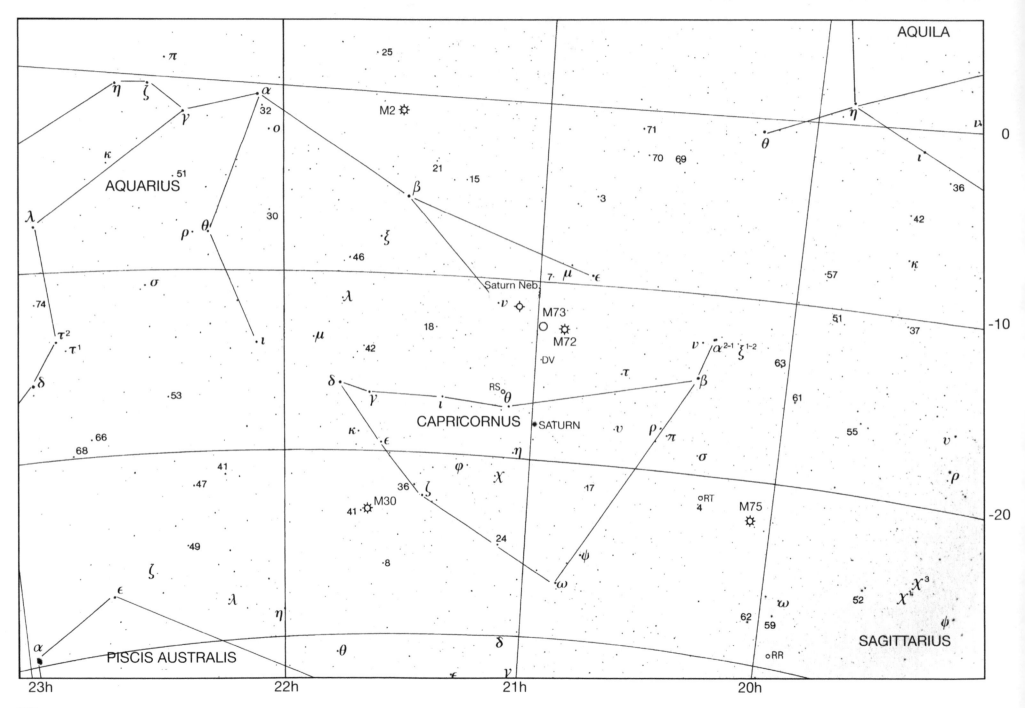

Map 45

PAVO, ARA, TELESCOPIUM

Here we have almost the whole of Pavo, together with Indus, Corona Australis, Ara, Telescopium and parts of Tucana, Grus, Sagittarius and Microscopium. Many of these have been described earlier; Indus with Map 9 and Corona Australis with Map 43. The little Southern Crown is excellently shown here, and note also the relative faintness of Alpha and the double Beta Sagittarii, notwithstanding the fact that they have been allotted the first two letters of the Greek alphabet. Alpha Gruis appears towards the left-hand edge of the photograph. This map includes no first-magnitude stars, though there are several of the second.

Pavo, one of the four Southern Birds (the others are Grus, Phœnix and Tucana) is not a very rich constellation, and its leader, Alpha (1.9, B3) is rather isolated from the main pattern; logically it seems that it ought really to belong to Indus! It is a powerful star, 700 times as luminous as the Sun and 230 light-years away. Next in order of brightness come Beta (3.4, A5), Delta (3.6, G5) and Eta (3.6, K1). Delta is of interest, because it is almost a twin of our Sun, and is less than 19 light-years away. Whether it has an Earth-like planet orbiting it we do not know, but there seems no logical reason why not. Close beside it on the map are Mu[1] (5.7, K0) and Mu[2] (5.3, also K0); they are certainly associated, but are a long way apart.

Kappa Pavonis is variable, with a range of from magnitude 3.9 to 4.7 and a period of just over 9 days. It used to be classed as a Cepheid, but is now known to belong to a sub-class known either as W Virginis variables (after the prototype star) or as Type II Cepheids. These variables are much less luminous than classical Cepheids

such as Eta Aquilæ and, of course, Delta Cephei itself; Kappa Pavonis is less than 4 times as luminous as the Sun, and is 75 light-years away. Useful comparison stars are Zeta Pavonis (4.0, K2; it is just off the bottom of this map), and Delta Aræ (3.6, B8). Do not use Lambda Pavonis, which is usually of just below the fourth magnitude, but is a Gamma Cassiopeiæ-type unstable star and can sometimes show outbursts raising it to magnitude 3.4. As with all these stars, its outbursts are quite unpredictable, and Lambda is worth monitoring with the naked eye.

S Pavonis is a semi-regular variable, one of a little triangle of stars below and to the right of Alpha; it is singled out by its redness. Below Gamma and Beta Pavonis on the map is a small semi-circle of stars, of which the brightest is the orange Omicron (5.1, M2); two of the other stars in the pattern, Y and SX, are red semi-regular variables.

NGC 6752 (C93) is a fine globular cluster, above and left of Lambda Pavonis and close to Omega (5.1, K2); on the photograph it appears in the guise of a somewhat bluish star. It is not far below naked-eye visibility, and in even a modest telescope it is a beautiful sight. Its distance is around 20,000 light-years. Below Lambda on the map is a group of stars of which Theta (5.9, A3) is the most prominent; above and to the left of this group is the galaxy NGC 6744 (C101), a relatively large though not bright barred spiral. It can just be identified on the photograph.

Telescopium adjoins Ara; its brightest star is Alpha (3.5, B3).

Note the colour of Alpha Tucanæ, well above Alpha Pavonis on the map. It is of type K, and its slightly

orange hue contrasts well with the bluish-white of Alpha Pavonis.

PAVO

Brightest stars

Star	Proper name	Mag.	Spectrum	RA h m s	Dec. ° ' "
α Alpha		1.94	B3	20 25 38.7	−56 44 06
β Beta		3.42	A5	20 44 57.4	−66 12 12
δ Delta		3.56	G5	20 08 43.3	−66 10 56
η Eta		3.62	K1	17 45 43.8	−64 43 25
κ Kappa		3.9 max.	F5v	18 56 56.9	−67 14 01
ε Epsilon		3.96	A0	20 00 35.4	−72 54 38

Variable stars

	RA h m	Dec. ° '	Range (mags.)	Type	Period (d)	Spectrum
λ	18 52.2	−62 11	3.4–4.3	Irreg.		B
κ	18 56.9	−67 14	3.9–4.7	W Virginis	9.09	F
T	19 50.7	−71 46	7.0–14.0	Mira	244	M
S	19 55.2	−59 12	6.6–10.4	Semi-reg.	386	M
Y	21 24.3	−69 44	5.7–8.5	Semi-reg.	233	N
SX	21 28.7	−69 30	5.4–6.0	Semi-reg.	50	M

Double star

	RA h m	Dec. ° '	PA °	Sep. sec	Mags.
ξ	18 23.2	−61 30	154	3.3	4.4, 8.6

Globular cluster

M/C	NGC	RA h m	Dec. ° '	Diameter min.	Mag.
C93	6752	19 10.9	−59 59	20.4	5.4

ARA

Brightest stars

Star	Proper name	Mag.	Spectrum	RA h m s	Dec. ° ' "
β Beta		2.85	K3	17 25 17.9	−55 31 47
α Alpha		2.95	B3	17 31 50.3	−49 52 34
ζ Zeta		3.13	K5	16 58 37.1	−55 59 24
γ Gamma		3.34	B1	17 25 23.5	−56 22 39
δ Delta		3.62	B8	17 31 05.8	−60 41 01
θ Theta		3.66	B1	18 06 37.6	−50 05 30
η Eta		3.76	K5	16 49 47.0	−59 02 29

Variable star

	RA h m	Dec. ° '	Range (mags.)	Type	Period (d)	Spectrum
R	16 39.7	−57 00	6.0–6.9	Algol	4.42	B

Open clusters

M/C	NGC	RA h m	Dec. ° '	Diameter min.	Mag.	No of stars
C82	6193	16 41.3	−48 46	15	5.2	
	6204	16 46.5	−47 01	5	8.2	45
	6208	16 49.5	−53 49	16	7.2	60
	6250	16 58.0	−45 48	8	5.9	60
	H13	17 05.4	−48 11	15		15
	IC 4651	17 24.7	−49 57	12	6.9	80

Globular clusters

M/C	NGC	RA h m	Dec. ° '	Diameter min.	Mag.
C81	6352	17 25.5	−48 25	7.1	8.1
	6362	17 31.9	−67 03	10.7	8.3
C86	6397	17 40.7	−53 40	25.7	5.6

TELESCOPIUM

Brightest star

Star	Proper name	Mag.	Spectrum	RA h m s	Dec. ° ' "
α Alpha		3.51	B3	18 26 58.2	−45 58 06

Variable stars

	RA h m	Dec. ° '	Range (mags.)	Type	Period (d)	Spectrum
RR	20 04.2	−55 43	6.5–16.5	Z And.		F5p
R	20 14.7	−46 58	7.6–14.8	Mira	461.9	M

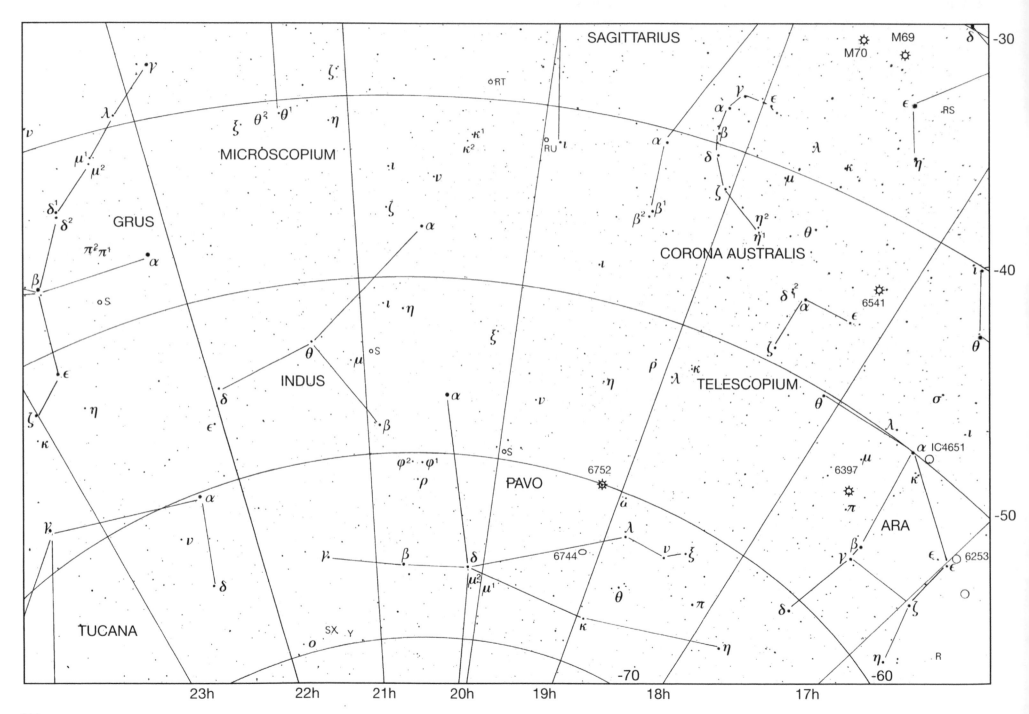

M	NGC/IC	Constellation	Type	RA		Dec.		Mag.	Size '	Notes
1	1952	Taurus	Supernova remnant	05	34.5	+22	01	8.4	6 × 4	Crab Nebula
2	7089	Aquarius	Globular cluster	21	33.5	−00	49	6.5	12.9	
3	5272	Canes Venatici	Globular cluster	13	42.2	+28	23	6.4	16.2	
4	6121	Scorpius	Globular cluster	16	23.6	−26	32	5.9	26.3	
5	5904	Serpens	Globular cluster	15	18.6	+02	05	5.8	17.4	
6	6405	Scorpius	Open cluster	17	40.1	−32	13	4.2	15	Butterfly
7	6475	Scorpius	Open cluster	17	53.9	−34	49	3.3	80	
8	6523	Sagittarius	Nebula	18	08.3	−24	23	6.0	90 × 40	Lagoon Nebula
9	6333	Ophiuchus	Globular cluster	17	19.2	−18	31	7.9	9.3	
10	6254	Ophiuchus	Globular cluster	16	57.1	−04	06	6.6	15.1	
11	6705	Scutum	Open cluster	18	51.1	−06	16	5.8	14	Wild Duck
12	6218	Ophiuchus	Globular cluster	16	47.2	−01	57	6.6	14.5	
13	6205	Hercules	Globular cluster	16	41.7	+36	28	5.9	16.6	
14	6402	Ophiuchus	Globular cluster	17	37.6	−03	15	7.6	11.7	
15	7078	Pegasus	Globular cluster	21	30.0	+12	10	6.3	12.3	
16	6611	Serpens	Nebula and cluster	18	18.8	−13	47	6.4	35 × 28	Eagle Nebula
17	6618	Sagittarius	Nebula	18	20.8	−16	11	7.0	46 × 37	Omega Nebula
18	6613	Sagittarius	Open cluster	18	19.9	−17	08	6.9	9	
19	6273	Ophiuchus	Globular cluster	17	02.6	−26	16	7.1	13.5	
20	6514	Sagittarius	Nebula	18	02.6	−23	02	7.6	29 × 27	Trifid Nebula
21	6531	Sagittarius	Open cluster	18	04.6	−22	30	5.9	13	
22	6656	Sagittarius	Globular cluster	18	36.4	−23	53	7.7	24.0	
23	6494	Sagittarius	Open cluster	17	56.8	−19	01	5.5	27	
24	–	Sagittarius	Star-cloud	18	16.9	−18	29	4.5	90	
25	IC 4725	Sagittarius	Open cluster	18	31.6	−19	15	4.6	32	Upsilon Sagittarii Cluster
26	6694	Scutum	Open cluster	18	45.2	−09	24	8.0	14	
27	6853	Vulpecula	Planetary nebula	19	59.6	+22	43	7.6	350" × 910"	Dumbbell Nebula
28	6626	Sagittarius	Globular cluster	18	24.5	−24	52	6.9	11.2	
29	6913	Cygnus	Open cluster	20	23.9	+38	32	6.6	7	
30	7099	Capricornus	Globular cluster	21	40.4	−23	11	7.5	11.0	
31	224	Andromeda	Sb galaxy	00	42.7	+41	16	3.5	178 × 63	Great Spiral
32	221	Andromeda	E2 galaxy	00	42.7	+40	52	8.2	7.6 × 5.8	Companion to M31
33	598	Triangulum	Sc galaxy	01	33.9	+30	39	5.7	62 × 39	
34	1039	Perseus	Open cluster	02	42.0	+42	47	5.2	35	
35	2168	Gemini	Open cluster	06	08.9	+24	20	5.0	28	
36	1060	Auriga	Open cluster	05	36.1	+34	08	6.0	12	
37	2099	Auriga	Open cluster	05	52.4	+32	33	5.6	24	
38	1912	Auriga	Open cluster	05	28.7	+35	50	6.4	21	
39	7092	Cygnus	Open cluster	21	32.2	+48	26	4.6	32	
40	–	–	Missing. Possibly a comet?							
41	2287	Canis Major	Open cluster	06	47.0	−20	44	4.5	38	
42	1976	Orion	Nebula	05	35.4	−05	27	5	66 × 60	Great Nebula
43	1982	Orion	Nebula	05	35.6	−05	16	7	20 × 15	Extension of M42

M	NGC/IC	Constellation	Type	RA		Dec.		Mag.	Size '	Notes
44	2632	Cancer	Open cluster	08	40.1	+19	59	3.1	95	Præsepe
45	1432/5	Taurus	Open cluster	03	47.0	+24	07	1.2	110	Pleiades
46	2437	Puppis	Open cluster	07	41.8	−14	49	6.1	27	
47	2422	Puppis	Open cluster	07	36.6	−14	30	4.4	30	
48	2548	Hydra	Open cluster	08	13.8	−05	48	5.8	54	
49	4472	Virgo	E4 galaxy	12	29.8	+08	00	8.4	8.9 × 7.4	
50	2323	Monoceros	Open cluster	07	03.2	−08	20	5.9	16	
51	5194	Canes Venatici	Sc galaxy	13	29.9	+47	12	8.4	11.0 × 7.8	Whirlpool
52	7654	Cassiopeia	Open cluster	23	24.2	+61	35	6.9	13	
53	5024	Coma Berenices	Globular cluster	13	12.9	+18	10	7.7	12.5	
54	6715	Sagittarius	Globular cluster	18	55.1	−30	29	7.7	9.1	
55	6809	Sagittarius	Globular cluster	19	40.0	−30	58	6.9	19.0	
56	6779	Lyra	Globular cluster	19	16.6	+30	11	8.2	7.1	
57	6720	Lyra	Planetary nebula	18	53.6	+33	02	9.7	70" × 150"	Ring Nebula
58	4579	Virgo	Sb galaxy	12	37.7	+11	49	9.8	5.4 × 4.4	
59	4621	Virgo	E3 galaxy	12	42.0	+11	39	9.8	5.1 × 3.4	
60	4649	Virgo	E1 galaxy	12	43.7	+11	33	8.8	7.2 × 6.2	
61	4303	Virgo	Sc galaxy	12	21.9	+04	28	9.7	6.0 × 5.5	
62	6266	Ophiuchus	Globular cluster	17	01.2	−30	07	6.6	14.1	
63	5055	Canes Venatici	Sb galaxy	13	15.8	+42	02	8.6	12.3 × 7.6	
64	4826	Coma Berenices	Sb galaxy	12	56.7	+21	41	8.5	9.3 × 5.4	Black-eye Galaxy
65	3623	Leo	Sb galaxy	11	18.9	+13	05	9.3	10.0 × 3.3	
66	3627	Leo	Sb galaxy	11	20.1	+12	59	9.0	8.7 × 4.4	
67	2682	Cancer	Open cluster	08	50.4	+11	49	6.9	30	
68	4590	Hydra	Globular cluster	12	39.5	−05	48	5.8	54	
69	6637	Sagittarius	Globular cluster	17	31.4	−32	21	7.7	7.1	
70	6681	Sagittarius	Globular cluster	18	43.2	−32	18	8.1	7.8	
71	6838	Sagitta	Globular cluster	19	53.8	+18	47	8.3	7.2	
72	6981	Aquarius	Globular cluster	20	53.5	−12	32	9.3	5.9	
73	6994	Aquarius	4 faint stars	20	58.9	−12	38	−	−	
74	628	Pisces	Sc galaxy	01	36.7	+15	47	9.2	10.2 × 9.5	
75	6684	Sagittarius	Globular cluster	20	06.1	−21	55	8.6	6.0	
76	650–1	Perseus	Planetary Nebula	01	42.5	+51	34	12.2	65" × 290"	Little Dumbbell
77	1068	Cetus	SBp galaxy	02	42.7	−00	01	8.8	6.9 × 5.9	
78	2068	Orion	Nebula	05	46.7	+00	03	10.0	8 × 6	
79	1904	Lepus	Globular cluster	05	24.5	−24	33	8.0	8.7	
80	6093	Scorpius	Globular cluster	16	17.0	−22	59	7.2	8.9	
81	3031	Ursa Major	Sb galaxy	09	55.6	+69	04	6.9	25.7 × 14.1	Bode's Galaxy
82	3034	Ursa Major	Pec. galaxy	09	55.8	+69	41	8.4	11.2 × 4.6	
83	5236	Hydra	Sc galaxy	13	37.0	−29	52	8.2	11.2 × 10.2	
84	4374	Virgo	E1 galaxy	12	25.1	+12	53	9.3	5.0 × 4.4	
85	4382	Virgo	S0 galaxy	12	22.9	+18	28	9.3	3	
86	4406	Virgo	E3 galaxy	12	25.2	+12	57	9.2	7.4 × 5.5	

M	NGC/IC	Constellation	Type	RA		Dec.		Mag.	Size '	Notes
87	4486	Virgo	E1 galaxy	12	30.8	+12	24	8.6	7.2 × 6.8	Virgo A
88	4501	Coma Berenices	SBb galaxy	12	32.0	+14	25	9.5	6.9 × 3.9	
89	4552	Virgo	E0 galaxy	12	35.7	+12	33	9.8	4.2	
90	4569	Virgo	Sb galaxy	12	36.8	+13	10	9.5	9.5 × 4.7	
91	–	–	Missing. Possibly a comet?							
92	6341	Hercules	Globular cluster	17	17.1	+43	08	6.5	11.2	
93	2447	Puppis	Open cluster	07	44.6	–23	52	6.2	22	
94	4736	Canes Venatici	Sb galaxy	12	50.9	+41	07	8.2	11.0 × 9.1	
95	3551	Leo	SBb galaxy	10	44.0	+11	42	9.7	7.4 × 5.1	
96	3368	Leo	Sb galaxy	10	46.8	+11	49	9.2	7.1 × 5.1	
97	3587	Ursa Major	Planetary nebula	11	14.8	+55	01	12.0	15".9	Owl Nebula
98	4192	Coma Berenices	Sb galaxy	12	13.8	+14	54	10.1	9.5 × 3.2	
99	4254	Coma Berenices	Sc galaxy	12	18.8	+14	25	9.8	5.4 × 4.8	
100	4321	Coma Berenices	Sc galaxy	12	22.9	+15	49	9.4	6.9 × 6.2	
101	5457	Ursa Major	Sc galaxy	14	03.2	+54	21	7.7	26.9 × 26.3	Pinwheel Galaxy
102	–	–	Missing. Possibly a comet, or else identical with M101?							
103	581	Cassiopeia	Open cluster	01	33.2	+60	42	7.4	25	
104	4594	Virgo	Sb galaxy	12	40.0	–11	37	8.3	8.9 × 4.1	Sombrero Galaxy
105	3379	Leo	E1 galaxy	10	47.8	+12	35	9.3	4.5 × 4.0	
106	4258	Canes Venatici	Sc galaxy	12	19.0	+47	18	8.3	18.2 × 7.9	
107	6171	Ophiuchus	Globular cluster	16	32.5	–13	03	8.1	10.0	
108	3556	Ursa Major	Sc galaxy	11	08.7	+55	57	10.0	8 × 1	
109	3992	Ursa Major	Sb galaxy	11	55.0	+53	39	9.5	7 × 4	
110	205	Andromeda	E6 galaxy	00	40.4	+41	41	8.0	17.4 × 9.8	Companion to M31

Messier's original catalogue ends with M103. The later numbers are galaxies discovered by Méchain. M104 was added to the catalogue by Flammarion in 1921, on the basis of finding a handwritten note about it in Messier's own copy of his 1781 catalogue. M105 and 107 were added by H.S. Hogg in the 1947 edition of the catalogue, and M108 and 109 by Owen Gingerich in 1960. NGC 205 has sometimes been referred to as M110, but this has not been generally accepted.

C	NGC/IC	Constellation	Type	RA		Dec.		Mag.	Size '	Notes
1	188	Cepheus	Open cluster	00	44.4	+85	20	8.1	14	
2	40	Cepheus	Planetary nebula	00	13.0	+72	32	11.6	0.6	
3	4236	Draco	Sb galaxy	12	16.7	+69	28	9.7	21 × 7	
4	7023	Cepheus	Nebula	21	01.8	+68	12	6.8	18 × 18	Bright reflection nebula
5	IC 342	Camelopardalis	SBc	03	46.8	+68	06	9.2	18 × 17	
6	6543	Draco	Planetary nebula	17	58.6	+66	38	8.8	0.3/5.8	Cat's Eye Nebula
7	2403	Camelopardalis	Sc galaxy	07	36.9	+65	36	8.9	18 × 10	
8	559	Cassiopeia	Open cluster	01	29.5	+63	18	9.5	4	
9	Sh2-155	Cepheus	Bright nebula	22	56.8	+62	37	7.7	50 × 10	Cave Nebula
10	663	Cassiopeia	Open cluster	01	46.0	+61	15	7.1	16	
11	7635	Cassiopeia	Bright nebula	23	20.7	+61	12	7.0	15 × 8	Bubble Nebula
12	6946	Cepheus	Sc galaxy	20	34.8	+60	09	9.7	11 × 9	
13	457	Cassiopeia	Open cluster	01	19.1	+58	20	6.4	13	Phi Cas Cluster
14	869/884	Perseus	Double cluster	02	20.0	+57	08	4.3	30 and 30	Sword Handle
15	6826	Cygnus	Planetary nebula	19	44.8	+50	31	9.8	0.5/2.3	Blinking Nebula
16	7243	Lacerta	Open cluster	22	15.3	+49	53	6.4	21	
17	147	Cassiopeia	dE4 galaxy	00	33.2	+48	30	9.3	13 × 8	
18	185	Cassiopeia	dE0 galaxy	00	39.0	+48	20	9.2	12 × 9	
19	IC 5146	Cygnus	Bright nebula	21	53.5	+47	16	10.0	12 × 12	Cocoon Nebula
20	7000	Cygnus	Bright nebula	20	58.8	+44	20	6.0	120 × 100	North America Nebula
21	4449	Canes Venatici	Irregular galaxy	12	28.2	+44	06	9.4	5 × 3	
22	7662	Andromeda	Planetary nebula	23	25.9	+42	33	9.2	0.3/2.2	
23	891	Andromeda	Sb galaxy	02	22.6	+42	21	9.9	14 × 2	
24	1275	Perseus	Seyfert galaxy	03	19.8	+41	31	11.6	2.6 × 1	Perseus A radio source
25	2419	Lynx	Globular cluster	07	38.1	+38	53	10.4	4.1	
26	4244	Canes Venatici	S galaxy	12	17.5	+37	49	10.6	16 × 2.5	
27	6888	Cygnus	Bright nebula	20	12.0	+38	21	7.5	20 × 10	Crescent Nebula
28	752	Andromeda	Open cluster	01	57.8	+37	41	5.7	50	
29	5005	Canes Venatici	Sb galaxy	13	10.9	+37	03	9.8	5.4 × 2	
30	7331	Pegasus	Sb galaxy	22	37.1	+34	25	9.5	11 × 4	
31	IC 405	Auriga	Bright nebula	05	16.2	+34	16	6.0	30 × 19	Flaming Star Nebula
32	4631	Canes Venatici	Sc galaxy	12	42.1	+32	32	9.3	15 × 3	
33	6992/5	Cygnus	Supernova remnant	20	56.4	+31	43	–	60 × 8	E. Veil
34	6960	Cygnus	Supernova remnant	20	45.7	+30	43	–	70 × 6	W. Veil
35	4889	Coma Berenices	E4 galaxy	13	00.1	+27	59	11.4	3 × 2	Brightest in Coma Cluster
36	4559	Coma Berenices	Sc galaxy	12	36.0	+27	58	9.8	10 × 4	
37	6885	Vulpecula	Open cluster	20	12.0	+26	29	5.7	7	
38	4565	Coma Berenices	Sb galaxy	12	36.3	+25	59	9.6	16 × 3	
39	2392	Gemini	Planetary nebula	07	29.2	+20	55	9.9	0.2/0.7	Eskimo Nebula
40	3626	Leo	Sb galaxy	11	20.1	+18	21	10.9	3 × 2	
41	–	Taurus	Open cluster	04	27.0	+16	00	1.0	330	Hyades
42	7006	Delphinus	Globular cluster	21	01.5	+16	11	10.6	2.8	Very distant globular
43	7814	Pegasus	Sb galaxy	00	03.3	+16	09	10.5	6 × 2	

C	NGC/IC	Constellation	Type	RA		Dec.		Mag.	Size '	Notes
44	7479	Pegasus	SBb galaxy	23	04.9	+12	19	11.0	4 × 3	
45	5248	Boötes	Sc galaxy	13	37.5	+08	53	10.2	6 × 4	
46	2261	Monoceros	Bright nebula	06	39.2	+08	44	10	2 × 1	Hubble's Variable Nebula
47	6934	Delphinus	Globular cluster	20	34.2	+07	24	8.9	5.9	
48	2775	Cancer	Sa galaxy	09	10.3	+07	02	10.3	4.5 × 3	
49	2237–9	Monoceros	Bright nebula	06	32.3	+05	03	–	80 × 60	Rosette Nebula
50	2244	Monoceros	Open cluster	06	32.4	+04	52	4.8	24	
51	IC 1613	Cetus	Irregular galaxy	01	04.8	+02	07	9.0	12 × 11	
52	4697	Virgo	E4 galaxy	12	48.6	−05	48	9.3	6 × 3	
53	3115	Sextans	E6 galaxy	10	05.2	−07	43	9.1	8 × 3	Spindle Galaxy
54	2506	Monoceros	Open cluster	08	00.2	−10	47	7.6	7	
55	7009	Aquarius	Planetary nebula	21	04.2	−11	22	8.3	2.5 × 1	Saturn Nebula
56	246	Cetus	Planetary nebula	00	47.0	−11	53	8.0	3.8	
57	6822	Sagittarius	Irregular galaxy	19	44.9	−14	48	9.3	10 × 9	Barnard's Galaxy
58	2360	Canis Major	Open cluster	07	17.8	−15	37	7.2	13	
59	3242	Hydra	Planetary nebula	10	24.8	−18	38	8.6	0.3/21	Ghost of Jupiter
60	4038	Corvus	Sc galaxy	12	01.9	−18	52	11.3	2.6 × 1.8	Antennæ
61	4039	Corvus	Sp galaxy	12	01.9	−18	53	13	3.2 × 2.2	Antennæ
62	247	Cetus	S galaxy	00	47.1	−20	46	8.9	20 × 7	
63	7293	Aquarius	Planetary nebula	22	29.6	−20	48	6.5	13	Helix Nebula
64	2362	Canis Major	Open cluster	07	18.8	−24	57	4.1	8	Tau CMa Cluster
65	253	Sculptor	Scp galaxy	00	47.6	−25	17	7.1	25 × 7	Sculptor Galaxy
66	5694	Hydra	Globular cluster	14	39.6	−26	32	10.2	3.6	
67	1097	Fornax	SBb galaxy	02	46.3	−30	17	9.2	9 × 6	
68	6729	Corona Australis	Bright nebula	19	01.9	−36	57	9.7	1.0	R CrA Nebula
69	6302	Scorpius	Planetary nebula	17	13.7	−37	06	12.8	0.8	Bug Nebula
70	300	Sculptor	Sd galaxy	00	54.9	−37	41	8.1	20 × 13	
71	2477	Puppis	Open cluster	07	52.3	−38	33	5.8	27	
72	55	Sculptor	SB galaxy	00	14.9	−39	11	8.2	32 × 6	Brightest in Sculptor Cluster
73	1851	Columba	Globular cluster	05	14.1	−40	03	7.3	11	
74	3132	Vela	Planetary nebula	10	07.7	−40	26	8.2	0.8	
75	6124	Scorpius	Open cluster	16	25.6	−40	40	5.8	29	
76	6231	Scorpius	Open cluster	16	54.0	−41	48	2.6	15	
77	5128	Centaurus	Peculiar radio galaxy	13	25.5	−43	01	7.0	18 × 14	Cen A radio source
78	6541	Corona Australis	Globular cluster	18	08.0	−43	42	6.6	13	
79	3201	Vela	Globular cluster	10	17.6	−46	25	6.7	18	
80	5139	Centaurus	Globular cluster	13	26.8	−47	29	3.6	36	Omega Centauri
81	6352	Ara	Globular cluster	17	25.5	−48	25	8.1	7	
82	6193	Ara	Open cluster	16	41.3	−48	46	5.2	15	
83	4945	Centaurus	SBc galaxy	13	05.4	−49	28	9.5	20 × 4	
84	5286	Centaurus	Globular cluster	13	46.4	−51	22	7.6	9	
85	IC 2391	Vela	Open cluster	08	40.2	−53	04	2.5	50	o Vel Cluster

C	NGC/IC	Constellation	Type	RA		Dec.		Mag.	Size '	Notes
86	6397	Ara	Globular cluster	17	40.7	−53	40	5.6	26	
87	1261	Horologium	Globular cluster	03	12.3	−55	13	8.4	7	
88	5823	Circinus	Open cluster	15	05.7	−55	36	7.9	10	
89	6067	Norma	Open cluster	16	18.9	−57	54	5.4	12	S Nor Cluster
90	2867	Carina	Planetary nebula	09	21.4	−58	19	9.7	0.2	
91	3532	Carina	Open cluster	11	06.4	−58	40	3.0	55	
92	3372	Carina	Bright nebula	10	43.8	−59	52	6.2	120 × 120	Eta Carinæ Nebula
93	6752	Pavo	Globular cluster	19	10.9	−59	59	5.4	20	
94	4755	Crux	Open cluster	12	53.6	−60	20	4.2	10	Jewel Box Cluster
95	6025	Triangulum Aus.	Open cluster	16	03.7	−60	30	5.1	12	
96	2516	Carina	Open cluster	07	58.3	−60	52	3.8	30	
97	3766	Centaurus	Open cluster	11	36.1	−61	37	5.3	12	
98	4609	Crux	Open cluster	12	42.3	−62	58	6.9	5	
99	–	Crux	Dark nebula	12	53.0	−63	00	–	400 × 300	Coal Sack
100	IC 2944	Centaurus	Open cluster	11	36.6	−63	02	4.5	15	Gamma Cen Cluster
101	6744	Pavo	SBb galaxy	19	09.8	−63	51	9.0	16 × 10	
102	IC 2602	Carina	Open cluster	10	43.2	−64	24	1.9	50	Theta Car Cluster
103	2070	Dorado	Bright nebula	05	38.7	−69	06	1.0	40 × 25	Tarantula Nebula
104	362	Tucana	Globular cluster	01	03.2	−70	51	6.6	13	
105	4833	Musca	Globular cluster	12	59.6	−70	53	7.3	14	
106	104	Tucana	Globular cluster	00	24.1	−72	05	4.0	31	47 Tucanæ
107	6101	Apus	Globular cluster	16	25.8	−72	12	9.3	11	
108	4372	Musca	Globular cluster	12	25.8	−72	40	7.8	19	
109	3195	Chamæleon	Planetary nebula	10	09.5	−80	52	–	0.6	

The Caldwell Catalogue was compiled in 1995 by one of the present authors (Moore); it contains objects readily accessible with small telescopes, but not listed by Messier. It was originally published in the periodical *Sky and Telescope*, and is now widely used. (Obviously the initial letter M could not be used, and the solution was to use the compiler's full surname, which is Caldwell-Moore!)

ABERRATIONS (OF LENSES) Small optical flaws are present in all photographic lenses and none are perfect. Thus a point of light can be reproduced as a small patch and a straight line as a slightly curved band. It is the task of lens designers (now largely dependent on computers) to control most aberrations by combining a number of single lenses or elements within a given lens such that the aberrations of one element will be cancelled out by the opposing aberrations of another element.

ABSOLUTE MAGNITUDE The apparent magnitude that a star would have if it could be seen from a standard distance of 10 parsecs (32.6 light-years).

ÅNGSTRÖM UNIT One hundred millionth part of a centimetre.

APERTURE The opening formed by the diaphragm inside a lens which can usually be varied. It controls the amount of light admitted for a given shutter speed, affects the depth of field (the zone of acceptable sharpness in front of and behind the prime subject) and also can reduce lens aberrations.

APPARENT MAGNITUDE The apparent brightness of a celestial body; the lower the magnitude, the brighter the object. The ancient Greeks initiated this system, the 20 brightest stars in the sky were "first magnitude", and the faintest that could be seen with the naked eye were "sixth magnitude". Each magnitude was about 2.5 times brighter than the next. This was later formalized so that the difference in brightness between different magnitude stars was 2.512, so that a first magnitude star is precisely 100 times as bright as a sixth magnitude star – 2.512 is the fifth root of 100, and the relationship is logarithmic.

ASTRONOMICAL UNIT (A.U.) The mean distance between the Earth and the Sun: 149,597,900 kilometres (92,958,400 miles).

ATMOSPHERE/"SEEING" The astronomical observer and astrophotographer have many problems to contend with but bad "seeing" is possibly the worst. Close to the ground heat currents can be generated by chimneys and other heating systems or by large areas of concrete. If it is feasible, the optimum siting of an observatory can deal with much of this. However, high altitude turbulence created by large-scale weather movements up to an altitude of about ten miles above the surface is much more difficult if not impossible to overcome. Light rays from planets or stars are refracted slightly each time they encounter bodies of air at different temperatures and this causes their image to blur or appear to wobble. A night full of "twinkling" stars may appear to be romantic but this twinkling or scintillation means that turbulence is marked and that any serious astronomical work is difficult to accomplish if not impossible. It must be said that these conditions are far more adverse for those using telescopes (whether for observation or imaging) where the images of objects are enlarged than when recording wide-angle images of the sky as in this book. The turbulence which causes bad seeing is not consistent and there are very brief periods when the skilled worker can see an object far more clearly and make the most of that opportunity to study or photograph it.

ATMOSPHERE/"TRANSPARENCY" This a measure of how bright celestial objects appear. The seeing might therefore be bad but the transparency good – as in the case of twinkling stars, see **Atmosphere/"seeing"**. Transparency can be affected by local conditions (for example, smoke generated in an industrial area) or by more general weather conditions such as haze or even dust and ash injected into the atmosphere by major volcanic eruptions. It can also be affected by normal, faint atmospheric processes such as airglow.

AZIMUTH The bearing of an object in the sky, measured from north (0°) through east, south and west.

BINARY SYSTEM A stellar system made up of two stars, genuinely associated and moving round their common centre of gravity.

CCD An acronym for charge-coupled device. This is an imaging chip containing thousands of pixels or picture elements. Incoming photons of light are converted into electrical charges which build up a picture that is stored in digital form and can be displayed on a monitor. The CCD is far more efficient than photographic film in converting the available light into the form of an image. Its response is linear so that – unlike film where efficiency falls off dramatically with the length of exposure – doubling the length of an exposure doubles the amount of information recorded. Its dynamic range (or latitude) is also superior and the production of data in digital form means that the computer can be used to process the information and to manipulate it. The major, professional observatories have been using CCD techniques for years and steady reductions in the price of chips has meant that amateurs can increasingly afford either to build their own CCD cameras or purchase them commercially ready-made. The area of the chips, however, is still very small so that they are totally unsuitable for wide field images of the type used in this book. In addition, a continuing weakness of the CCD compared with photographic film under optimum conditions is its lower resolution – or ability to discern fine detail.

CELESTIAL SPHERE An imaginary sphere surrounding the Earth's globe, whose centre coincides with the centre of the Earth. The celestial poles lie above the Earth's poles, and the celestial equator is a projection of the Earth's equator on the celestial sphere. The celestial co-ordinates **Right Ascension** and **Declination** specify the position of celestial bodies.

CLUSTER VARIABLES (OR CLUSTER-CEPHEIDS) An obsolete name for the stars now known as RR Lyræ variables.

COLOUR BALANCE The average, overall colour in a transparency, print or other form of reproduction. In an everyday scene we would expect to see, for example, skin tones, trees and the sky appearing approximately as we see them in reality. Departures from that expectation might result from one or more of the hues not being reproduced accurately or being under- or oversaturated. In a wide-angle astronomical image the most critical feature is the manner in which the darkness of the sky is reproduced. Any departure from a neutral black or blue-black is regarded as undesirable. Any general departure of this kind from a norm in an image is usually referred to as a colour cast. (See also **Density** below.)

COLOUR INDEX The difference between a star's visual magnitude and its photographic magnitude. The redder the star, the greater the positive value of the colour index. Bluish stars have negative colour indices. For stars of spectral type A0, the colour index is zero.

COMA This is a lens aberration in which point sources of light near the edges of an image frame are reproduced as comet-shaped blurs (hence the name) with the tails pointing towards the centre of the image. It is a particular problem with wide-angle lenses featuring large maximum apertures although stopping down a lens improves the condition. Aspherical lenses – those whose curved surfaces, unlike most lenses, do not conform to the shape of a sphere – assure maximum correction of coma but they are considerably more expensive.

CONSTELLATION A group of stars named after a living or mythological character, or an inanimate object. Eighty-eight constellations are recognized by the International Astronomical Union.

DECLINATION (DEC.) The angular distance of a celestial body north or south of the celestial equator expressed in degrees (°) and minutes ('). It is equivalent to the latitude measurement on Earth.

DENSITY This is usually given a mathematical definition in photography but it may be regarded as a measure of the effect of light upon film or paper. The extent of density in exposed and developed film will depend upon the amount of light that has fallen upon it. When using negative black and white film, for example, overexposure will result in very dense (black) areas and underexposure in thin or much less dense areas. The reverse is true with transparency films: underexposure results in dark or dense images and over exposure in thin washed out positive images. The ideal, of course, is an image with a full range of density levels. A vital piece of information about the performance of a film is its characteristic curve – a graph which plots density against exposure. One aspect of density is very important when film is used in astrophotography – and particularly so in wide-field photography of the type presented in this book where large areas of sky are included. A processed transparency film exhibits its maximum density (normally referred to as D-Max) in those areas that have been totally unaffected by light – the borders of frames and the ends of the film roll. The density within the frame when photographing a good, dark sky will be somewhat less but still very considerable. Now, when photographic scientists design a colour film they concentrate on the best possible rendering of individual colours and such features as skin tones. When combined together in a photograph the colours may not integrate to a neutral black D-Max in the areas of less exposure – the shadows. This does not matter in everyday photography because we do not usually examine shadows very closely. But in wide-field images of the constellations, for example, any colour bias in a film's D-Max is important because a very high percentage of the image area occupied by the sky will be devoted to it. Slight bias can be tolerated but occasionally otherwise very successful films have to be rejected because of an objectionable D-Max colour cast.

DIRECT MOTION Movement of revolution or rotation in the same sense as that of the Earth.

DISTORTION A lens aberration which alters the shape of objects in a picture, not their sharpness. It is not corrected by stopping down and takes two forms – barrel distortion in which straight lines are bowed in at the edges of a frame resembling the sides of a barrel and the opposite – pincushion distortion – in which straight lines are bowed in towards the middle of a frame. The former problem tends to be present to some degree in wide-angle lenses and the latter in small amounts in telephoto lenses.

DOUBLE STAR A star made up of two or more components, either genuinely associated (binary system) or merely in the same line of sight (optical pair).

DRIVE (OR TRACKING) PLATFORM A unit on which a camera is locked into place to take long exposures of celestial objects and which, by means of an electric motor or motors, compensates for the rotation of the Earth by driving in the opposite direction. Even with relatively wide-angle lenses it needs to be accurately aligned on the north celestial pole (in the northern hemisphere). Cameras can of course be "piggy backed" on telescopes whilst dedicated camera drive units can range from the very simple to the far more complex – an example of the latter being a unit which not only permits operation at sidereal, lunar and solar tracking rates but which can be programmed to track the movement of individual comets.

DWARF NOVÆ A term sometimes applied to SS Cygni (U Geminorum) stars.

DWARF STAR A dwarf star is one which lies on the **Main Sequence** but is too small to be classified as a **Giant** or Supergiant. Red Dwarf stars are stars of **Spectral Type** K or M; White Dwarfs are the remnants of Red Giants once nuclear fusion ceases, and they in turn become Black Dwarfs once they have radiated all their remaining energy. Brown Dwarfs are potential stars, but with too little mass to cause fusion and to ignite. These theoretical objects have yet to be detected.

ECLIPSING BINARY (OR ECLIPSING VARIABLE) A binary system in which one component is regularly occulted or hidden by the other, so that the light which we receive from the star is reduced. The prototype eclipsing binary is Algol.

ED This is a proprietary term standing for Extra-low Dispersion glass used in lenses. Different wavelengths of light are bent by different amounts when they pass through normal optical glass and, since the rays would not be brought to focus at the same plane, correcting this chromatic aberration as it is termed is another task for lens designers. Such correction is generally effective with normal and wide-angle lenses but the longer focal length, telephoto lenses magnify even the slightest difference in focus between light rays. The special ED glass was developed to meet this added complication and to maintain contrast and sharpness even at maximum apertures.

EMULSION Technically, the suspension of light sensitive silver salts in a gelatin solution to form a coating on photographic film and papers. Not infrequently, however, it is used by photographers as another expression for "film" – particularly when comparing one film with another.

EPOCH A date chosen for reference purposes in quoting astronomical data. The epoch chosen for this Atlas is 2000.0

EQUATOR, CELESTIAL The projection of the Earth's equator on to the celestial sphere.

EXTENDED OBJECTS A term used to describe virtually all celestial objects other than stars which are point sources of light. Thus the Moon is a bright extended object whilst nebulæ or comets are much dimmer extended objects. There is a major difference between photographing extended objects and point sources which is dealt with in the item below – **Speed (of a Lens)**.

F-NUMBER A numerical expression of the relative aperture of a lens at its different diaphragm openings. A lens will typically have the f-numbers 2.8, 4, 5.6, 8, 11, 16 and 22 engraved on it and these represent the focal length of the lens divided by the effective aperture of the lens opening at that point. The lower the f-number the more light is admitted (by opening the aperture) and the higher the number the less light falls on the film during an exposure. Each of the numbers indicates a halving or a doubling of the amount of light admitted relative to the f-numbers on either side. Thus compared to an f-number of 5.6, selecting an effective aperture of 4 will double the amount of light admitted while selecting 8 will halve it. These values are usually presented as f/4, f/5.6, f/8 and so on. Basically, all lenses set to the same f-number produce images of equal illumination and this is one of the major factors on which guidance on exposure determination is calculated. The f-number also enables the actual diameter of a chosen lens's aperture to be worked. Thus a 50mm f/1.4 lens at its maximum aperture has a diameter of 35.7mm (50mm divided by 1.4) but set to f/2.8 the diameter becomes 17.9mm.

F-STOP The term stop drawn from early photographic equipment is now used interchangeably with f-number. Stop is also used in the expression to stop down and refers to reducing the affective aperture of a lens to control aberrations or to reduce the amount of light reaching the film.

FILM – PRODUCTION BATCH Film manufacturers – like almost all manufacturers – have a complex schedule of production runs. A major effort is made to match the characteristics of any one film in its various production runs but there may well be slight differences which might be noticeable when the processed film is examined critically. It is a general practice in professional and technical photography to secure supplies of film from a single batch in those uses where slight differences could adversely affect a successful outcome.

FOCAL LENGTH Technically, the distance from the rear nodal point of a lens to the film plane when focused at infinity. Expressed in terms of millimetres, the shorter the focal length the wider the picture angle and the smaller the image size – whilst the longer the focal length, the narrower the picture angle and the larger the image size. Thus used with a 35mm camera, a wide-angle 28mm focal length lens covers an angular field of view of 46° in the vertical by 65° in the horizontal. A 200mm focal length telephoto lens covers 7° by 10°. In popular terms, a short focal length lens will include more and reduce the size of any individual element within the frame: a longer focal length lens will appear to "close in" on a scene and thus appear to increase the size of any element within the frame.

GALAXIES Systems of stars, nebulæ and interstellar matter.

GIANT STAR Any star with a radius greater than 10 times that of our Sun. Blue Giants lie on the **Main Sequence**; a Red Giant is a stars that has depleted its hydrogen content, is burning helium, and expands dramatically. Once fusion ceases altogether, it collapses as a White **Dwarf**, or if its mass is sufficiently great, it will explode as **Supernova**.

HERTZSPRUNG-RUSSELL (H-R) DIAGRAM A diagram in which stars are plotted to their **Spectral Type** and **Absolute Magnitude**. As stars evolve they change their position on the diagram. Stable, hydrogen-burnign stars are clustered on the **Main Sequence**, a curving line from the top left of the diagram to the bottom right. Once the hydrogen is exhausted, a star leaves the Main Sequence.

ISO Stands for International Standards Organisation and an ISO number attached to a film is an official rating of its speed or sensitivity to light. The various ISO speeds have now replaced the previously used ASA (American Standards Association) ratings although in fact they have exactly the same numerical values. ISO speed ratings are determined in a strictly regulated, standard manner. In this they differ from the increasingly used EI (Exposure Index) system which advances different speed ratings for a film depending on – for example, light source, subject matter, desired effect and developer used. The use of EI received a powerful stimulus a decade ago with the appearance of Kodak's T-Max films (followed by other leading manufacturers) – which may be regarded as new technology products which could be "tuned" to different speeds with minimum adverse effects compared with older generations of film.

LENS: TELEPHOTO A lens whose focal length is longer than the diagonal of the film frame which in 35mm photography means lenses longer than 50mm or 58mm. Often referred to simply as a long lens.

LENS: WIDE-ANGLE Usually defined as a lens whose focal length is shorter than the diagonal of the film frame. In 35mm photography, this means in effect a lens of shorter focal length than the "standard" or "normal" lens which has a focal length of 50mm. (The latter is regarded as normal because its angle of view is about 45°, which is similar to that portion of human vision in which we discern sharp detail.)

LIGHT POLLUTION Broad term applied to the effect of urban lighting on the night sky. In the absence of moonlight and of lighting generated by largely urban development, the sky is extremely dark and very faint objects such as nebulæ and galaxies can be seen with the naked eye. Extensive street lighting, advertising lighting and security lights especially if they are badly designed throw light upwards as well as downwards. The resulting light scatter renders dim objects invisible and in very bad cases leaves only the brightest of stars for the city dweller to see. In badly light-polluted areas all but the shortest of photographic exposures yield poor results with dark sky and celestial objects being swamped by colour casts generated by lighting. The much shorter exposures generally required by the CCD and the ability to readily manipulate images subsequently have been stressed by proponents as making imaging from light polluted areas far more feasible using CCDs than when film is used. In recent years a growing international campaign by the astronomical community to persuade lighting engineers, communities and authorities to install well designed lighting which directs light downwards where it is wanted has had its effect but the growth of urbanization generally and reluctance to change existing systems makes the struggle a hard one.

LIGHT-YEAR The distance travelled by light in one year: 9.4607 million million km, or 5.88 million million miles.

LOCAL GROUP A group of over two dozen galaxies, of which our Galaxy is a member. Its largest member is the Andromeda Spiral M31.

MAIN SEQUENCE A band in the H-R Diagram from top

left to lower right, containing most of the stars – apart from the giants and the white dwarfs. Our Sun is a typical Main Sequence star.

NEBULA A cloud of dust and gas in space, inside which fresh stars may be formed.

NEUTRON STAR A remnant of a very massive star produced after a supernova explosion. Neutron stars send out radio pulses, and are therefore known as pulsars.

NOVA A star which temporarily flares up to many times its normal brilliancy. The outburst happens in the white dwarf component of a binary system.

PARALLAX, TRIGONOMETRICAL The apparent shift of an object when observed from two different directions.

PARSEC The distance at which a star would show a parallax of one second of arc: 3.26 light-years, 206,265 astronomical units, or 30.857 million million kilometres (19.174 million million miles).

PLANETARY NEBULA A small, hot, dense highly-evolved star which has thrown off its outer layers, and is surrounded by a shell of gas. All planetary nebulæ are expanding.

POSITION ANGLE The apparent direction of one object with reference to another, measured from north (0°) through east, south and west.

PRECESSION The apparent slow motion of the celestial poles. This also means a shift of the celestial equator, and hence of the equinoxes; the vernal equinox moves by 50 seconds of arc per year, and has moved out of Aries into Pisces.

PROPER MOTION The individual motion of a star against the background stars.

RIGHT ASCENSION (RA) The angular distance of a celestial body from the vernal equinox – the point at which the ecliptic crosses the projection of the Earth's equator on the celestial sphere – measured eastwards in hours (h) and (m). It is equivalent to the longitude measurement on Earth.

SKY FOG This is not the normal form of fog but a phrase used to describe the cumulative effects of various forms of light in the sky which can build up during time exposures so that the light from the fainter stars is swamped before it can be recorded. Light pollution may be included in this consideration but the term is usually applied to natural phenomena such as faint auroræ, the zodiacal light, airglow (for example, the interaction between solar radiation and atomic oxygen in the upper atmosphere) and even scattered sunlight during summer nights. It should not be confused with fog as a photographic term, which is the density found inherently in an unexposed but developed film. That fog level can be increased by various factors not attributable to image forming activity.

SLR – SINGLE LENS REFLEX The most popular current camera for the serious amateur photographer (when using 35mm film) and for many professional photographers in both 35mm and medium format (for example, 6×6cm) sizes. The term "reflex" refers to a system whereby the photographer views the scene through the lens being used by means of a 45° angled mirror which reflects the image upwards to a ground glass viewing screen. The term single lens reflex was introduced to differentiate this type of camera from cameras which use two lenses of equal focal length – the top one being used for viewing and focusing via a mirror.

SLR may be regarded as something of a misnomer, however, because one of the most important features of this type of camera has traditionally been the availability of numerous lenses of different focal lengths that can be used. "Multiple lens reflex" might perhaps have been a more accurate description.

SPECTRAL TYPE Stars are classified according to the appearance of the absorption lines in their spectra. The classes run OBAFGKM representing different colours, with subdivisions 0–9.

SPEED (OF A LENS) The maximum f-number or focal ratio of a lens is a basic indication of its speed capability when photographing everyday objects or extended objects in astrophotography. Thus an f/4.5 lens would now be regarded as quite slow and an f/2.8 lens as average. Many keen amateurs would doubtless aim for an f/1.4 lens. There are both advantages (use of a high shutter speed) and disadvantages (probable lens aberrations) in using a fast lens wide open but the new astrophotographer needs to understand a significant difference between judging a lens's speed when photographing everyday objects or extended celestial objects and when photographing stars. Stars are point sources of light and it is not the focal ratio of a lens but its absolute diameter that indicates how fast or slow it is. Thus an everyday photographer would regard a 35mm f/1.4 lens as being faster than a 180mm f/2.8 lens when used wide open and this is correct – in ordinary photography and shooting the Moon or a nebula. But when photographing stars, the long lens is the faster. The diameter of the wide angle lens is 25mm (35/1.4) but that of the longer lens 64mm (180/2.8). Therefore if both lenses were used wide open, with the same speed of film and for a similar length of time, the 180mm lens would record a greater number of fainter stars than would the 50mm lens.

SPEED (OF FILM) A term used to describe the sensitivity of film to light (see ISO above). Today films of up to about ISO100 are regarded as being of slow speed; those from IS0100 up to about 400 medium speed; and any above that as being high speed. In the last category speeds of ISO1000, 1600, 3200 and even 6400 are by no means uncommon. Years ago slow speed films could be described as offering by far the best quality with fast films offering little more than the ability to secure some sort of image of subjects photographed under poor lighting conditions. Modern film technology has introduced very significant improvements in this direction, however, and the general quality obtained with the medium speed and some of the high speed films is such that it can sometimes be difficult, when looking at the results, to decide exactly what speed of film was used. Nonetheless, it is still generally sound advice – and certainly so in astrophotography – that the photographer should use a lower speed film rather than a higher speed film where appropriate for the end result.

SUPERNOVA A colossal stellar outburst.
(Type I) The total destruction of the white dwarf component of a binary system.
(Type II) The collapse of a very massive star.

VARIABLE STAR Any star whose luminosity varies either as a result of internal processes or because of external phenomena such as being eclipsed by another star.

WOLF-RAYET STARS Very hot stars which are surrounded by expanding gaseous envelopes. Their spectra show bright emission lines.

ZENITH The observer's overhead point (altitude 90°).

ZODIAC A belt stretching round the sky, 8 degrees to either side of the ecliptic, in which the Sun, Moon and principal planets are always to be found. (Pluto can leave the Zodiac, and of course many asteroids do so.)

ASTROPHOTOGRAPHY
Astrophotography Basics, Publication P-150, Eastman Kodak, Rochester NY
Arnold HJP, *Astrophotography: An Introduction*, George Philip, London, 1995
Covington M, *Astrophotography for the Amateur*, CUP, 1991
Dragesco J, *High Resolution Astrophotography*, CUP, 1995
Dobbins TA, Parker DC,& Capen CF, *Observing and Photographing the Solar System*, Willmann-Bell, 1988
Gordon B, *Astrophotography*, Willmann-Bell, 1983
Malin D, *A View of the Universe*, CUP, 1993
Malin D, & Murdin P, *Colours of the Stars*, CUP, 1984
Martinez P, *Astrophotography II*, Wilmann-Bell, 1987
Wallis BD, & Provin RW, *A Manual of Advanced Celestial Photography*, CUP, 1988

ASTRONOMY
Burnham, R, *Burnham's Celestial Handbook*, 3 vols, Dover Books, 1978
Doherty, P, *Atlas of the Planets*, Hamlyn, 1985
Hartung, EJ, *Astronomical Objects for Southern Telescopes*, CUP, 1968
Jones, KG, *Messier's Clusters and Nebulæ*, Faber, 1970
Laustsen, S, Madsen, C & West, RM, *Exploring the Southern Sky*, Springer Verlag, 1987
Malin, D, *A View of the Universe*, CUP, 1993
Moore, P, *The Guinness Book of Astronomy*, Guinness, 1995
Webb, TW, *Celestial Objects for Common Telescopes*, Dover Books, 1972 (a new reprint of this famous classic)
The *Yearbook of Astronomy* is published annually by Macmillan

Proper names are given in italics, followed by each star's official annotation and the name of its constellation.

Acamar	θ	Eridanus
Achernar	α	Eridanus
Achird	η	Cassiopeia
Acrux	α	Crux Australis
Adhafera	ζ	Leo
Adhara	ε	Canis Major
Agena	β	Centaurus
Ain	ε	Taurus (Hyades)
Al Dhanab	β	Grus
Al Giedi	α²	Capricornus
Al Kaprah	κ	Ursa Major
Al Nair al Kentaurus	ζ	Centaurus
Al Nath	β	Taurus
Al Rischa	α	Pisces
Al Suhail al Wazn	λ	Vela
Alava	η	Serpens (*Cauda*)
Albaldah	π	Sagittarius
Albali	ε	Aquarius
Albireo	β	Cygnus
Alcyone	η	Taurus (Pleiades)
Aldebaran	α	Taurus
Alderamin	α	Cepheus
Aldhibah	ζ	Draco
Aldhibain	η	Draco
Algenib	γ	Pegasus
Algieba	γ	Leo
Algjebbah	η	Orion
Algol	β	Perseus

Algorel	δ	Corvus
Alheka	ζ	Taurus
Alhena	γ	Gemini
Alioth	ε	Ursa Major
Alkaffaljidhina	γ	Cetus
Alkafzah	χ	Ursa Major
Alkaid	η	Ursa Major
Almaak	γ	Andromeda
Almaaz	ε	Auriga
Alnair	α	Grus
Alnasr	γ	Sagittarius
Alnilam	ε	Orion
Alnitak	ζ	Orion
Alniyat	σ	Scorpius
Alphard	α	Hydra
Alphekka	α	Corona Borealis
Alpheratz	α	Andromeda
Alpherg	η	Pisces
Alphirk	β	Cepheus
Alrai	γ	Cepheus
Alshain	β	Aquila
Altair	α	Aquila
Altarf	β	Cancer
Althalimain	λ	Aquila
Alula Australis	ξ	Ursa Major
Alula Borealis	ν	Ursa Major
Alwaid	β	Draco
Alya	θ	Serpens (*Cauda*)
Alzirr	ξ	Gemini
Angetenar	τ⁴	Eridanus
Ankaa	α	Phœnix
Antares	α	Scorpius

Arcturus	α	Boötes
Arich	γ	Virgo
Arkab	β	Sagittarius
Arneb	α	Lepus
Asad Australis	ε	Leo
Ascella	ζ	Sagittarius
Asellus Australis	δ	Cancer
Asmidiske	ξ	Puppis
Ati	o	Perseus
Atik	ζ	Perseus
Atlas	27	Taurus (Pleiades)
Atria	α	Triangulum Australe
Avior	ε	Carina
Azha	η	Eridanus
Baten Kaitos	ζ	Cetus
Baten Kaitos Shemali	ι	Cetus
Bellatrix	γ	Orion
Betelgeux	α	Orion
Biham	θ	Pegasus
Canopus	α	Carina
Capella	α	Auriga
Castor	α	Gemini
Chaph	β	Cassiopeia
Cheleb	β	Ophiuchus
Chort	θ	Leo
Cor Caroli	α²	Canes Venatici
Dabih	β	Capricornus
Deneb	α	Cygnus
Deneb al Giedi	δ	Capricornus
Denebola	β	Leo
Dheneb	ζ	Aquila
Diphda	β	Cetus

Dschubba	δ	Scorpius
Dubhe	α	Ursa Major
Edasich	ι	Draco
Electra	17	Taurus (Pleiades)
Eltamin	γ	Draco
Enif	ε	Pegasus
Fomalhaut	α	Piscis Australis
Garnet Star	μ	Cepheus
Giansar	λ	Draco
Gienah	ε	Cygnus
Girtab	κ	Scorpius
Gomeisa	β	Canis Minor
Gorgonea Terti	ρ	Perseus
Graffias	β	Scorpius
Hamal	α	Aries
Han	ζ	Ophiuchus
Haratan	θ	Centaurus
Hassaleh	ι	Auriga
Hatysa	ι	Orion
Heka	λ	Orion
Heze	ζ	Virgo
Homan	ζ	Pegasus
Hyadum Primus	γ	Taurus (Hyades)
Izar	ε	Boötes
Jabbah	ν	Scorpius
Jabhat al Akrab	ω[1]	Scorpius
Juza	ξ	Draco
Kaus Australis	ε	Sagittarius
Kaus Borealis	λ	Sagittarius
Kaus Meridionalis	δ	Sagittarius
Ke Kwan	κ	Centaurus
KeKouan	β	Lupus
Kerb	τ	Perseus
Kitalpha	α	Equuleus
Kocab	β	Ursa Minor
Koo She	δ	Vela
Kornepheros	β	Hercules
Kraz	β	Corvus
Kursa	β	Eridanus
Lesath	υ	Scorpius
Maia	20	Taurus (Pleiades)
Marfik	λ	Ophiuchus
Markab	α	Pegasus
Markeb	κ	Vela
Matar	η	Pegasus
Mebsuta	ε	Gemini
Megrez	δ	Ursa Major
Mekbuda	ζ	Gemini
Men	α	Lupus
Menkar	α	Cetus
Menkarlina	β	Auriga
Menkent	γ	Centaurus
Merak	β	Ursa Major
Mesartim	γ	Aries
Miaplacidus	β	Carina
Minelauva	δ	Virgo
Minkar	γ	Corvus
Mintaka	δ	Orion
Mirach	β	Andromeda
Miram	η	Perseus
Mirphak	α	Perseus
Mirzam	β	Canis Major
Misam	κ	Perseus
Mizar	ζ	Ursa Major
Muscida	o	Ursa Major
Nair al Butain	41	Aries
Nashira	γ	Capricornus
Nekkar	β	Boötes
Nihal	β	Lepus
Nunki	σ	Sagittarius
Nusakan	β	Corona Borealis
Persian	α	Indus
Phad	γ	Ursa Major
Phakt	α	Columba
Pherkad Major	γ	Ursa Minor
Phurad	ζ	Canis Major
Polaris	α	Ursa Minor
Polis	μ	Sagittarius
Pollux	β	Gemini
Præcipua	46	Leo Minor
Procyon	α	Canis Minor
Propus	η	Gemini
Rana	δ	Eridanus
Rasalgethi	α	Hercules
Rasalhague	α	Ophiuchus
Rasalmothallah	α	Triangulum
Rassalas	μ	Leo
Regor	γ	Vela
Regulus	α	Leo
Rigel	β	Orion
Rijl al Awwa	μ	Virgo
Rotanev	β	Delphinus
Ruchbah	δ	Cassiopeia
Rukbat	α	Sagittarius
Rutilicus	ζ	Hercules
Sabik	η	Ophiuchus

T - #0906 - 101024 - C220 - 210/290/10 - PB - 9780750306546 - Gloss Lamination